New Materials for Next-Generation Commercial Transports

Committee on New Materials for Advanced Civil Aircraft

National Materials Advisory Board
Aeronautics and Space Engineering Board

Commission on Engineering and Technical Systems

National Research Council

Publication NMAB-476
NATIONAL ACADEMY PRESS
Washington, D.C. 1996

NOTICE: The project that is the subject of this report was approved by the Governing Board of the National Research Council, whose members are drawn from the councils of the National Academy of Sciences, the National Academy of Engineering, and the Institute of Medicine. The members of the committee responsible for the report were chosen for their special competences and with regard for appropriate balance.

This report has been reviewed by a group other than the authors according to procedures approved by a Report Review Committee consisting of members of the National Academy of Sciences, the National Academy of Engineering, and the Institute of Medicine.

The National Academy of Sciences is a private, nonprofit, self-perpetuating society of distinguished scholars engaged in scientific and engineering research, dedicated to the furtherance of science and technology and to their use for the general welfare. Upon the authority of the charter granted to it by the Congress in 1863, the Academy has a mandate that requires it to advise the federal government on scientific and technical matters. Dr. Bruce M. Alberts is president of the National Academy of Sciences.

The National Academy of Engineering was established in 1964, under the charter of the National Academy of Sciences, as a parallel organization of outstanding engineers. It is autonomous in its administration and in the selection of its members, sharing with the National Academy of Sciences the responsibility for advising the federal government. The National Academy of Engineering also sponsors engineering programs aimed at meeting national needs, encourages education and research, and recognizes the superior achievements of engineers. Dr. Harold Liebowitz is president of the National Academy of Engineering.

The Institute of Medicine was established in 1970 by the National Academy of Sciences to secure the services of eminent members of appropriate professions in the examination of policy matters pertaining to the health of the public. The Institute acts under the responsibility given to the National Academy of Sciences by its congressional charter to be an adviser to the federal government and, upon its own initiative, to identify issues of medical care, research, and education. Dr. Kenneth I. Shine is president of the Institute of Medicine.

The National Research Council was organized by the National Academy of Sciences in 1916 to associate the broad community of science and technology with the Academy's purposes of furthering knowledge and advising the federal government. Functioning in accordance with general policies determined by the Academy, the Council has become the principal operating agency of both the National Academy of Sciences and the National Academy of Engineering in providing services to the government, the public, and the scientific and engineering communities. The Council is administered jointly by both Academies and the Institute of Medicine. Dr. Bruce M. Alberts and Dr. Harold Liebowitz are chairman and vice chairman, respectively, of the National Research Council.

This study by the National Materials Advisory Board and the Aeronautics and Space Engineering Board was conducted under Grant No. FAA-93-G-040 with the U.S. Department of Transportation.

Library of Congress Catalog Card Number 96-67749
International Standard Book Number 0-309-05390-0

Available in limited supply from:
National Materials Advisory Board
2101 Constitution Avenue, NW
HA-262
Washington, D.C. 20418
202-334-3505

Additional copies are available for sale from:
National Academy Press
2101 Constitution Avenue, NW
Box 285
Washington, D.C. 20055
800-624-6242 or 202-334-3313 (in the Washington metropolitan area)

Printed in the United States of America.

COMMITTEE ON NEW MATERIALS FOR ADVANCED CIVIL AIRCRAFT

JOHN A.S. GREEN (Chair), Lockheed Martin Laboratories, Baltimore, Maryland
BERNARD BUDIANSKY, Harvard University, Cambridge, Massachusetts
DAVID J. CHELLMAN, Lockheed Martin Aeronautical Systems Company, Marietta, Georgia
LARRY P. CLARK, Boeing Defense and Space Group, Seattle, Washington
JOHN W. GILLESPIE, JR., University of Delaware, Newark
CHARLES E. HARRIS, NASA Langley Research Center, Hampton, Virginia
MURRAY H. KUPERMAN, United Airlines Maintenance and Operations Center, San Francisco, California
PAUL A. LAGACE, Massachusetts Institute of Technology, Cambridge
VICKI E. PANHUISE, AlliedSignal Aerospace, Tempe, Arizona
KENNETH L. REIFSNIDER, Virginia Polytechnic Institute and State University, Blacksburg
MICHAEL P. RENIERI, McDonnell Douglas Aerospace, St. Louis, Missouri
EDGAR A. STARKE, University of Virginia, Charlottesville
HERBERT J. WARDELL, Gulfstream Aerospace, Savannah, Georgia

Aeronautics and Space Engineering Board Liaison Representative

C. JULIAN MAY, Tech/Ops International, Inc., Kennesaw, Georgia

National Materials Advisory Board Staff

THOMAS E. MUNNS, Senior Program Officer
AIDA C. NEEL, Senior Project Assistant
JACK HUGHES, Research Associate

Aeronautics and Space Engineering Board Staff

ALAN C. ANGLEMAN, Senior Program Officer

National Materials Advisory Board

Aeronautics and Space Engineering Board

Acknowledgments

The recommendations and conclusions of this report are the insights of many people and organizations with whom the committee interacted. In particular, the efforts of the following individuals who presented detailed briefings to the committee are greatly appreciated: Douglas Cairns (manager, Advanced Composites Technology, Hercules) on innovative composite processing (processing/performance relationships); Robert Crowe (Defense Science Office, Advanced Research Projects Agency) on smart materials and structures; Robert Gog (technical representative, Lufthansa Airlines) on aircraft operations and maintenance; Gerald Janicki (director of Advanced Materials and Structures, McDonnell Douglas Aerospace–Transport Aircraft) on new materials applications for subsonic aircraft; Donald E. Larsen (Advanced Technology Division, Howmet Corporation) on net-shape investment cast titanium components for application in civil aircraft; J.A. Marceau (Materials Technology, Boeing Commercial Airplane Group) on service experience with current (aging) fleet; S.G. Sampath (FAA Technical Center) on aging aircraft issues related to current commercial fleet lessons learned; Brian W. Smith (program unit chief, Materials Technology, Boeing Commercial Airplane Group) on market driven materials development and engineering needs in materials and processes for commercial airplanes; Darrel Tenney (Materials Division) and James Starnes (Structures Division, NASA Langley Research Center) on new materials and structures projected for advanced subsonic aircraft.

The support and encouragement of the Federal Aviation Administration are greatly appreciated; the committee also thanks Pramode Bhaghat, Peter Shyprykevich, Joseph Soderquist, and Bill Wall who participated in numerous meetings.

The committee especially acknowledges the efforts of Thomas Munns, senior program officer for the National Materials Advisory Board, and Alan Angleman, senior program officer for the Aeronautics and Space Engineering Board, who maintained the continuity of the study and provided all necessary staff support. They were ably assisted by Aida Neel and Jack Hughes.

Preface

Turbulence is normally a term associated with flying. However, in the recent past, the airline industry seems to have experienced a great deal of turbulence on the ground as well. Airlines have been buffeted by a combination of forces following deregulation. First, there was the broad and lengthy recession, followed by numerous fare wars, the intense competition from newer airlines, and most recently, the concern over the safety of commuter flights. All these forces have seriously impacted the financial health of the airline industry as a whole. In fact, the competition has become so intense that some well-known airlines are now struggling for survival. This situation, in turn, has influenced the aircraft manufacturers who, in response, have adopted a pragmatic "no-frills" approach toward future design and manufacturing developments.

It is against this dynamic background that the Committee on New Materials for Advanced Civil Aircraft embarked on this study concerning the application of new materials in the next generation of subsonic transports on behalf of the Federal Aviation Administration. It is with considerable trepidation that one approaches a task of such complexity, attempting to project technical developments within the industry up to 15–20 years in the future, when even near-term materials and structures developments seem unpredictable. However, after extensive debate, the committee believes there are some clear paths along which technology will evolve. What is more difficult to predict in this turbulent era is the timing of these developments.

The committee was highly interactive, working best with lively debate and discussion. It brought together a good balance of industrial and academic expertise, along with government experience, particularly from the National Aeronautics and Space Administration. There was a balance on the committee between experts knowledgeable in advanced metallic materials and organic matrix composites. In addition, committee expertise covered the entire spectrum of materials use, from the innovation of new materials; to alloy and composite selection, fabrication, design and manufacturing; to in-service experience and nondestructive evaluation and maintenance. Experience related to smaller executive aircraft as well as large transports was also represented on the committee.

In addition to drawing upon their own sources of information, members of the committee elected to use a series of indepth, expert briefings to focus discussion on key areas of materials research, development, manufacturing, and application. In these briefings, aircraft manufacturing and maintenance experts, material and component suppliers, industry and government research leaders with both commercial and military experience, and materials and structures researchers helped formulate the recommendations and conclusions of this report. Their insights added greatly to the scope of this report.

The purpose of this study is to identify engineering issues related to the introduction of new materials and their expected effect on the life-cycle durability of next-generation commercial transports. The committee investigated the likely new materials and structural concepts for the next-generation commercial aircraft and the key factors influencing application decisions. Based on these predictions, the committee identified and analyzed the design, characterization, monitoring, and maintenance issues that appeared to be most critical for the introduction of advanced materials and structural concepts. The scope of this study did not include issues related to the High-Speed Civil Transport or the hot stages of turbine engines, although the ancillary components of engines that may become warm in application (e.g., thrust reversers) are included. Also considered outside the scope of the study were specific issues related to rotorcraft. Accordingly, the primary focus of the committee was defined as the identification of new materials and structures for the category of large subsonic transport aircraft; the general aviation category was also included where there were related problems or concerns.

It is now our belief that, despite the prevailing (and probably continuing) turbulence in the airline industry, this report should provide some insight into the evolution of advanced materials and processing technology on next-generation commercial aircraft. In doing so, it will provide information that can help to maintain a safe, efficient, and viable commercial fleet.

Comments or suggestions that readers of this report wish to make can be sent via Internet electronic mail to nmab@nas.edu or by FAX to the National Materials Advisory Board (202) 334-3718.

John A.S. Green, *Chair*
Committee on New Materials
for Advanced Civil Aircraft

Contents

Tables and Figures

TABLES

FIGURES

Executive Summary

Aircraft safety is the overriding concern of the flying public and is the primary design criteria for the aircraft industry. While the aircraft industry enjoys an excellent safety record, the continued increase in the number of passengers and the size of the commercial transport fleet requires significant reductions in accident rates to achieve the Federal Aviation Administration (FAA) goal of 50 percent reduction in fatalities by the turn of the century (FAA, 1991).

In response to the Aviation Safety Research Act of 1988 (P.L. 100-591), the FAA has initiated a comprehensive, long-range research program concerned with all aspects of civil aircraft safety. The overall objective of the FAA program is to anticipate the technological advances that will likely affect future aircraft and to evaluate their implications for aircraft safety.

New aircraft (i.e., aircraft that, in the context of this report, will enter service in the next 15–20 years) will incorporate new materials, fabrication processes, and structural concepts to realize the cost and performance benefits of lighter-weight, more-efficient structures. One very important concern regarding aircraft safety is the degradation of structural materials as aircraft operate beyond their original design life, bringing to the forefront issues of materials stability, corrosion resistance, fatigue behavior, and maintenance procedures. Evaluation of new materials and structures for commercial aircraft application requires an assessment of performance and operating costs over the entire life cycle of the aircraft, from fabrication to maintenance. Also, a thorough understanding of the material properties and anticipated performance in airline service is required to adequately develop inspection and maintenance procedures that assure safety over the service life of the aircraft.

The major objective of this study was to identify issues related to the introduction of new materials and the effect that advanced materials will have on the durability and technical risk of future civil aircraft throughout their service life. The committee investigated the new materials and structural concepts that are likely to be incorporated into next-generation commercial aircraft and the factors influencing application decisions. Based on these predictions, the committee attempted to identify the design, characterization, monitoring, and maintenance issues that are critical for the introduction of advanced materials and structural concepts into future aircraft. This study focuses primarily on airframe structures of large subsonic commercial transport aircraft with the understanding that many of the issues identified will also apply to smaller general aviation planes (both fixed and rotary wing). The emphasis of this study is, for the most part, restricted to primary and secondary airframe structures.

FINDINGS

There has been significant progress in the introduction of new materials and structural designs on commercial aircraft. Aircraft designers continue to apply new materials and structural concepts to provide benefits in performance, durability, compliance with environmental regulations, and most recently, acquisition and maintenance costs. The committee believes that the use of new materials and structures will continue to expand on next-generation aircraft.

The current, turbulent economic climate affecting the airline, manufacturer, and materials industries has significantly changed the application criteria for advanced materials. As a result, aircraft manufacturers are responding to airline concerns about reducing overall costs including the costs of acquisition and maintenance. The result is incremental, evolutionary—rather than revolutionary—changes in materials. The principal barriers to increased use of new high-performance materials are acquisition, manufacturing, certification, and life-cycle costs; incomplete understanding of failure mechanisms and their interactions; technological risk; and the state of the materials supplier base.

Principal "new" airframe materials expected to see increased use in the next generation of advanced civil aircraft include polymer-matrix composites (laminates, tailored forms, woven and sewn three-dimensional configurations, automated tape and tow placement) and metallic alloys (tough aluminum, high-yield-strength aluminum, aluminum-lithium, high-strength titanium, high-strength steel, and cast products). In contrast, given the emphasis on incremental technological advances and total costs, the committee does not foresee significant application of metal-matrix composites in the airframes of next-generation transports.

Increasingly, airframe manufacturers are using an integrated product development approach that considers such factors as producibility, cost, nondestructive evaluation (NDE) methods and criteria, and repair and maintenance issues; and involves airline designers, manufacturers, and suppliers from the outset of development programs.

Commercial aircraft are built and operated on a global basis with international teaming of manufacturers, suppliers, and fabricators. Accordingly, the development and harmonization of international standards for materials and processes, testing and evaluation, NDE, and repair and maintenance procedures is critical to developing and commercializing new materials and structures technology. In spite of the international teaming involved in developing a new aircraft, the manufacture and operation of commercial transports remains an extremely competitive business. The committee believes that competitive pressures will continue to influence the selection criteria for the application of new materials and processing technology.

RECOMMENDATIONS

Technological advances are brought about through the concerted efforts of airline, industry, academic, and government organizations. In forming their recommendations, the committee identified the technologies that are likely to be involved in the development of next-generation aircraft and outlined the work required to bring about those developments. The recommendations are directed toward all of the organizations involved in new materials applications, but also specifically toward what the FAA role should be in these developments.

- In general, the committee recommends that the FAA remain involved in all stages of the technology development process, with emphasis on work related to aircraft safety, operations, maintenance, and nondestructive evaluation.

Materials, Manufacturing, and Structural Concepts

Improvements in aircraft structural components will continue to be based on factors related to materials selection, analytical methods, structural concepts, and processing innovations. With the increased emphasis on affordability, it is likely that fewer new materials will be developed. On the other hand, robust and cost-effective processing methods as well as compliance with environmental regulations will become paramount issues to provide lower costs. Specifically, the committee envisions continued improvements in the performance and durability of metal alloys and polymeric composites. Materials and structures more conducive to low-cost processes such as casting, high-speed machining, and superplastic forming/diffusion bonding of metals and fiber placement, resin transfer molding, and nonautoclave processing of composites will be emphasized in the future.

The committee recommends that the FAA work with industry, government, and academic organizations in the development of new materials, processing, and structure technology by the following guidelines:

- Keep abreast of innovative materials processing technologies that provide methods for low-cost fabrication of aircraft structure. Emphasis should be placed on the understanding of new product forms, processing methods, and thermal treatments and their possible effects on materials performance.
- Support the development of emerging process modeling techniques for definition of processing parameters and requirements.
- Establish and maintain databases of material and structural properties resulting from the candidate processing methods. The databases should include test methods, physical and mechanical properties, failure modes, and influences of probable defects and manufacturing processes on property behavior.
- Work with the materials, manufacturing, and airline industries to develop industrywide standards to improve consistency in the final products, especially with the increasing globalization of materials availability.
- Participate in industry- and NASA-sponsored flight hardware demonstration programs for the introduction of new materials, manufacturing processes, and structural concepts in high-risk applications. FAA emphasis should be on validation of inspection and repair techniques and in the development of technology needed to certify and monitor these structures.

Methodologies for Assessment of Structural Performance

Current structural design and analytical procedures used by the aircraft industry are largely semiempirical, even though significant improvements have occurred in structural analysis methods over the last two decades. Accurate, finite element analysis methods are routinely used to predict the stress, strain, and displacement fields in complex structural geometries. However, the reliable prediction of structural failure modes, ultimate strength, residual strength, and fatigue life have remained elusive to the structural engineer. The current standard practice relies heavily on extensive testing at the coupon, subelement, element, subcomponent, component, and full-scale levels. Design details are frequently optimized through test programs. Scale-up effects are handled through a building-block approach that relies on testing to verify the anticipated structural performance at each scale level. While the committee anticipates that this building-block approach to structural design will continue indefinitely, a more rigorous, analytical prediction methodology will greatly improve the process of introducing new materials into airframe primary structure.

The committee recommends that the FAA work with other industry, government, academic organizations in the development of improved analytical methods by the following guidelines:

- Support development and facilitate implementation of advanced analytic and computational methodology to predict residual strength as a function of time.
- Support programs to improve the understanding of basic failure mechanisms in advanced materials and their structures. Include the interactions of the various failure modes manifested at the various length scales—from material to structural levels.

Inspection, Maintenance, and Repair

The successful application of new materials and structural concepts relies on an effective maintenance program that is cost-effective while ensuring passenger safety. The aging aircraft experience has provided the airline industry with significant lessons learned for inspection and repair technologies. These lessons provide a framework for improving inspection and repair processes for next-generation materials and structures. Major issues that continue to limit the effectiveness of an aircraft maintenance program are poor structural inspection standards, inadequate defect indication interpretation, unreliable inspection techniques, high cost of new NDE methods, and limited linkage with design analyses and NDE results. The leadership of the FAA and the continued participation of airlines and manufacturers in developing and implementing improved maintenance and inspection methods is crucial.

The committee recommends that the FAA take a leadership position in the development of improved inspection and maintenance methods by the following guidelines:

- Support the development of improved standards for NDE methodologies and their specific materials and structural applications, especially through participation in industry- and NASA-sponsored component development and flight hardware demonstration programs for the introduction of new materials, manufacturing processes, and structural concepts.
- Support the development of cost-effective, quantitative NDE methodologies for in-service inspection of airframe materials and structures. Emphasize improved defect detection reliability, cost-effectiveness, and ease of implementation in field environments. Particular attention should be given to rapid, wide-area inspection with limited or one-sided access.
- Develop improved analytic methods to determine NDE reliability and inspectability of materials and structures to support damage tolerance and durability analyses.
- Support the development of real-time repair and maintenance processes for materials and structures that make use of the results from quantitative NDE methods and computational analyses.

REPORT ORGANIZATION

The findings of the committee have been grouped into five parts. The first provides an overview of the emerging trends and the requirements and drivers for new materials applications. This overview is followed with separate discussions on materials, processes, and structural concepts; analytical methods; and aircraft operations. Specific conclusions and recommendations that emerged as a consensus view during the deliberations of the committee are cited in part 5.

I

Introduction

1

Requirements and Drivers

BACKGROUND

The Aviation Safety Research Act of 1988 (P.L. 100-591) directed the Federal Aviation Administration (FAA) to conduct research to develop technology to assess and improve aircraft safety. The law stipulates that the FAA spend at least 15 percent of its research budget on long-term investigations. In response to the Act, the FAA has initiated a comprehensive research program concerned with all aspects of civil aircraft safety. The overall objective of the FAA program is to anticipate the technological advances that will likely affect future aircraft and to evaluate their implications for aircraft safety.

One very important concern regarding aircraft safety is the degradation of structural materials as aircraft operate beyond their original design life, bringing to the forefront issues of materials stability, corrosion resistance, fatigue behavior, and maintenance procedures. Industry and government, led by the FAA, are actively addressing aging of the existing commercial aircraft fleet. Clearly, it will be important to exploit and disseminate the lessons learned from the experience of the existing fleet.

New aircraft (i.e., aircraft that, in the context of this report, will enter service in the next 15–20 years) will incorporate new materials, fabrication processes, and structural concepts to realize the cost and performance benefits of lighter-weight, more-efficient structures. Use of new structural materials, such as reinforced polymeric composites and advanced light-weight metallic alloys, is increasing as performance requirements become more demanding and service environments more severe.

Evaluation of new materials and structures for commercial aircraft application requires an assessment of performance and operating costs over the entire life cycle of the aircraft, from fabrication to maintenance. Also, a thorough understanding of the long-term material properties and anticipated failure mechanisms is required to adequately develop inspection and maintenance procedures that assure safety over the service life of the aircraft.

The major objective of this study was to identify issues and potential research and technology development opportunities related to the introduction of new materials on the next generation of commercial transports and the expected effect on the life-cycle durability. The committee investigated the likely new materials and structural concepts for next-generation commercial aircraft and the key factors influencing application decisions. Based on these predictions, the committee identified and analyzed the design, characterization, monitoring, and maintenance issues that appeared to be most critical for the introduction of advanced materials and structural concepts. The committee then developed recommendations for needed research aimed at addressing knowledge gaps. Methods by which the FAA could leverage its research efforts with related research in industrial, academic, and other governmental laboratories are also suggested. Evaluation of the FAA's established role in the certification of commercial aircraft was not within the scope of this study, however the recommended research should provide the FAA with an improved understanding of new materials and structures technology to aid in future certification activities.

This study focuses primarily on airframe structures of large subsonic commercial aircraft. However, many of the issues identified will also apply to smaller, general aviation planes (both fixed and rotary wing). The emphasis of this study is, for the most part, restricted to primary and secondary airframe structures.[1] While beyond the scope of this study, new materials technology could also significantly benefit major aircraft subsystems, including engines, brakes, auxiliary power units, and environmental control units.

INDUSTRY OVERVIEW AND IMPORTANT TRENDS

To identify likely applications of new materials and structures in future commercial aircraft, it is critical to understand the factors that influence materials selection and design decisions. New materials and structures applications depend on the types of aircraft that will be built in the future and the performance requirements for those aircraft. The current economic health of the airline industry and future market analyses will dictate the criteria for new aircraft.

[1] Aircraft structural components are considered to be *primary structure* if catastrophic failure of the component could lead to the loss of the aircraft. Failure of *secondary structural* components do not lead to the loss of the aircraft.

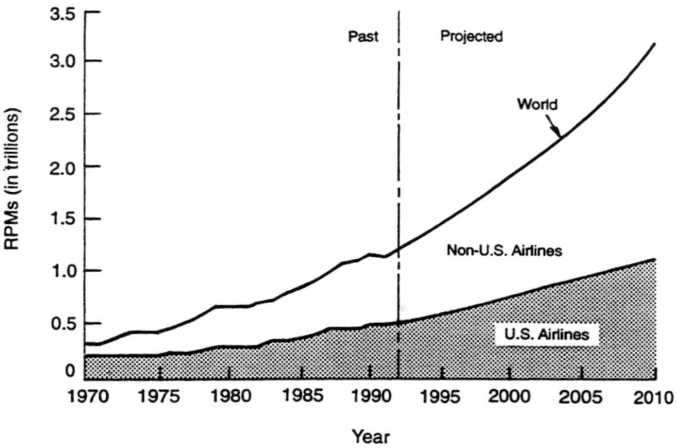

FIGURE 1-1 Past and projected trends in world air travel in revenue-passenger miles (RPM). Source: Janicki (1994).

Air travel, in terms of revenue-passenger miles (RPM),[2] has steadily increased over the past 20 years (figure 1-1). This trend is expected to continue well into the next century. More passengers are expected to travel more miles, with routes in the Pacific Rim seeing the greatest increase (ATA, 1993). Recently however, increased travel has not translated into greater profits for the airlines. The world's airlines lost more money (approximately $9 billion) in the years 1990–1993 than they had made since the end of World War II (ATA, 1993, 1995). Figure 1-2 shows the variability during the past decade in annual profits and losses for the airline industry.[3] Even with the steady increase in the number of people traveling, the costs associated with aircraft purchase, operation, and maintenance, coupled with a worldwide economic recession and fierce competition resulting from the deregulation of the U.S. airline industry, the viability of many of the largest airlines is threatened.

As a result of the airline industry's financial troubles, the market for new aircraft has been weak. The cost of acquisition of new advanced aircraft and relatively low fuel prices have been driving carriers to either retain their aircraft for a longer period of time than originally planned or to purchase (or lease) older or refurbished aircraft from other carriers. Therefore, one of the primary sources of competition for manufacturers of new aircraft are existing or refurbished aircraft. While the basic performance (e.g., number of seats and range) of a new airplane is of considerable importance, the most important airline market factor is the purchase and operating costs incurred by the airlines (Smith, 1994b; Janicki, 1994).

As shown in figure 1-3, aircraft manufacturers forecast a sizable market for new aircraft, based on the projections of substantial growth in passenger travel requirements and the replacement of aging aircraft in the current fleet. While the airline industry has been showing signs of recovering from the economic difficulties of the past five years (ATA, 1995), the recent trends described above will have a profound effect on the design and production of next-generation commercial aircraft.

NEXT-GENERATION AIRCRAFT

The major aircraft manufacturers have recently introduced new aircraft, including the Airbus A340, the McDonnell Douglas MD-11, and most recently, the Boeing 777. New aircraft models range from derivative models that are modifications of existing aircraft designs (e.g., fuselage extensions for increased capacity) to major redesigns with new wings, engines, and empennage. Derivative models have the advantage of similarity to an existing model, leading to reduced design and certification costs. Another benefit is that the similarity of operation and maintenance of derivative models to the existing fleet leads to greater acceptance by the airlines.

Due to the current poor market conditions for any new aircraft, coupled with the high costs and business risks involved with developing new aircraft models, the introduction of additional all-new aircraft in the near term (before the turn of the century) is unlikely. However, derivative aircraft production will continue, and these aircraft frequently exploit the benefits of new materials.

Aircraft manufacturers are currently investigating the feasibility and market for a very large commercial transport (VLCT). The current VLCT configurations under study have double-deck passenger cabins with a 500–800 seating capacity and a range of approximately 6,000–8,000 miles (AWST, 1994a,b,c). The large diameter fuselage, the large material stock sizes required, and the weight limitations imposed by runway loading restrictions in the development of such an

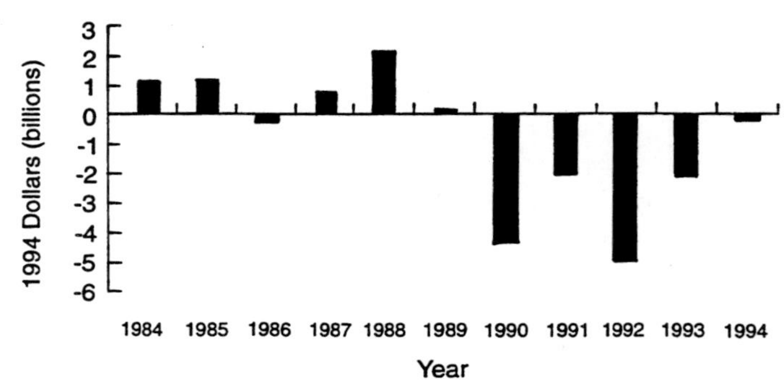

FIGURE 1-2 After-tax profits and losses (based on 1994 dollars) for U.S. scheduled airlines. Data Source: ATA (1995).

[2] A revenue-passenger mile is a measure of air traffic used by the airline industry and is defined as one fare-paying passenger transported one mile (ATA, 1995).

[3] Based on the results of the first three quarters of 1995, U.S. airlines are expected to have after-tax profits for the first time since 1989.

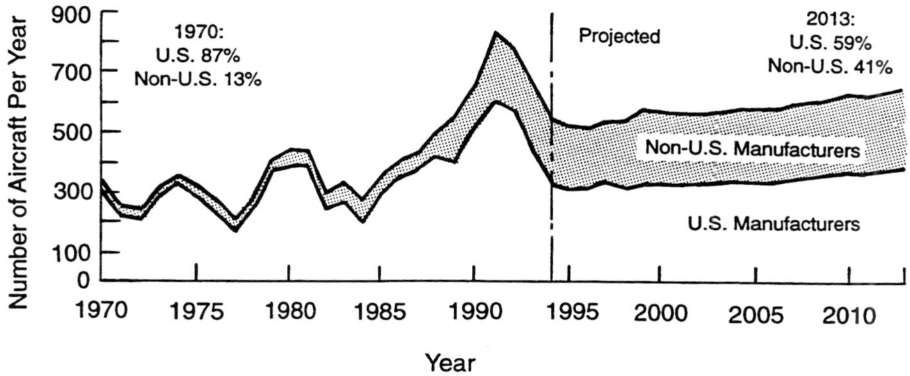

FIGURE 1-3 Past and projected manufacturing of commercial transports. Source: Janicki (1994).

aircraft would require significant development in technology and manufacturing capability.

APPLICATION OF NEW MATERIALS

Drivers and Barriers

The potential for the application of new materials exists on all new aircraft, including current production, derivatives, and new models. Drivers in the development and application of new materials traditionally have included performance (e.g., weight and range), durability, and regulatory compliance (e.g., environmental issues). Recently, the cost of acquisition and maintenance has become an overriding concern to the airlines. The result is that all new materials must demonstrate acceptable cost and sufficient performance improvements to justify or "pay their way" into production application.

Airframe life-cycle costs, including ownership costs (airplane price and financing expenses) and operating costs (fuel and maintenance), are a significant part of the total cost of an aircraft to an airline. Figure 1-4 shows a breakdown of direct operating costs based on extremes in fuel prices. The largest contributor to airline costs is the cost of ownership. As shown in figure 1-5 the airframe represents a significant fraction (37 percent) of the total cost of the aircraft. Regardless of aviation fuel prices, the costs related to airframe materials and structures are significant components of the total costs. The trends in environmental regulations worldwide will make end-of-life disposal or recycling of components more important factors in life-cycle costs.

Another concern indirectly related to the application of new materials is the health of the advanced materials industry itself. The economic viability of a large sector of the advanced materials industry is threatened because of significant reductions in the military markets, the economic decline in commercial aerospace, unrealistic market forecasts, and increased international competition (Seagle, 1994). In the current business environment, it is difficult for materials suppliers to commit to expensive design, development, and qualification efforts needed to evaluate and fabricate a new material for commercial aircraft applications. The necessary long lead times, significant resource commitments, and perceived high

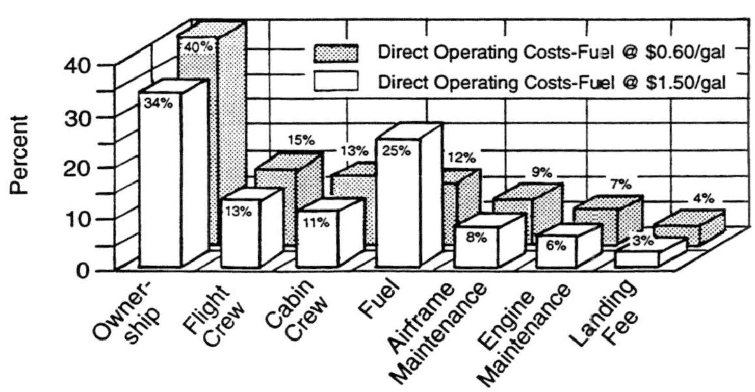

FIGURE 1-4 Breakdown of direct operating costs for two fuel-price estimates. Source: Janicki (1994).

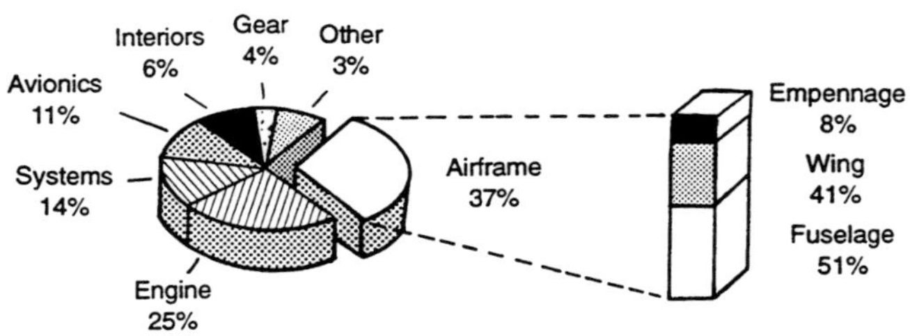

FIGURE 1-5 Breakdown of total aircraft cost. Source: Janicki (1994).

technical risks involved in competing for applications represent major barriers for materials suppliers.

In the past, the commercial aircraft industry has taken advantage of materials that were developed and characterized as part of military programs. Significant progress, most noticeably in the development and understanding of structural composites and significant production capability in advanced materials, has come through military research and development and hardware development efforts. As a result of recent reductions in U.S. Department of Defense acquisition programs, the benefit to the commercial industry from military programs will be largely curtailed. Nevertheless, the materials and aircraft industry must seek ways to continue to develop improved materials and innovative processes.

For current-production aircraft, the incentives to introduce new materials include improved maintenance and reliability, conformance with new environmental regulations, and manufacturing cost reduction. With the possible exception of environmental compliance, material substitutions on current aircraft will only be made if there is minimal impact on designs or tooling. The high cost of material qualification, changes to engineering drawings and data sets, and component certification are significant barriers to materials substitutions.

For derivative or new aircraft, the incentive to introduce new materials include reduced airline maintenance and operating costs, reduced production and rework costs, reduced manufacturing cycle time, and compliance with environmental regulations. Although qualification costs can inhibit the introduction of new materials and structural concepts on derivative models, the consideration of incentives described above are evolving production toward short development cycles, cost reductions, and customer-preferred options. Weight savings can also be a significant factor on new or derivative designs, especially for long-range aircraft. As a design matures, the issue of weight can become increasingly important in order to meet performance commitments. Barriers to new materials applications on new and derivative aircraft include the cost of developing an engineering and processing database and manufacturing capability and the general reluctance of airlines to accept the higher potential cost and risk in the operation and maintenance of airplanes due to new materials and structural concepts.

Aircraft safety remains a central concern of the aircraft industry and the fundamental basis for structural design criteria. Because of the excellent safety record held by commercial aviation, the public is less willing to accept aircraft accidents as inevitable (AWST, 1995).

Airlines and manufacturers are conservative about implementing new technologies because they are less willing to accept business risks (including liability) introduced by the technological risks associated with new technologies. The implementation of new materials always represents a technological risk that must be weighed against the performance benefits and the confidence in design and evaluation procedures.

Projected Evolutionary Advances

As in the past, new materials and structural concepts will inevitably but gradually be introduced into commercial aircraft. During the next 20 years, the projected technological readiness of new materials will result in their utilization in primary structures that include:

- advanced metallic fuselage,
- structural composite wing, and
- composite fuselage components.

The specific timing of these advances is very difficult to project because of the turbulent business climate and the dynamic nature of the factors influencing aircraft design.

Aircraft manufacturers have recently employed design teams and concurrent engineering approaches to implement new materials and structures technologies (Smith, 1994a). By involving all of the major stakeholders—including airline customers and design engineering, manufacturing, and supplier organizations—in the early design stages, their requirements can be addressed early in the development of component design. For example, airline concerns over operational and maintenance costs influence initial design criteria, component producibility becomes a key criterion, supplier-provided materials characteristics can be effectively tailored to needs, and known materials limitations can be accommodated during the first component design. This approach has significantly increased the production application of new developmental materials. For example, on the Boeing 777, approximately 60 percent of the program's materials development efforts that were underway when the decision was made to produce the airplane led to production applications (Smith, 1994a), an impressive achievement from a developmental viewpoint.

II

Materials, Processes, and Structural Concepts

The application of new materials and structures technology in commercial transport aircraft depends on a number of interrelated considerations. This part of the report describes trends and forecasts for technology development in materials technology, manufacturing processes, and innovative structural concepts and is organized in four chapters:

- Chapter 2, "Structural Concepts," discusses key design and technology issues in the development of innovative structural concepts.
- Chapter 3, "Metallic Materials and Processes," identifies development and trends in metal alloy materials and process development.
- Chapter 4, "Polymeric Composite Materials and Processes," provides an overview of key composite material and process technologies for next-generation transports.
- Chapter 5, "Environmentally Compliant Materials and Processes," describes the effect of environmental factors on the design, production, and operation of commercial transports.

FACTORS INFLUENCING MATERIAL SELECTION

Aircraft designers continue to apply new materials and structural concepts to provide benefits in performance, durability, compliance with environmental regulations, and most recently, acquisition and maintenance costs. The committee believes that the use of new materials and structures will continue to expand for next-generation aircraft.

The evaluation, selection, and introduction of new materials for aircraft applications is an involved process. Generally, it is accomplished through a series of decision points designed to minimize technical risk by identifying issues and mitigation strategies as early as possible in the development process.

Developing and characterizing a new material represents a large commitment of resources and time for both the materials supplier and the manufacturer. New materials often make improved structural concepts feasible. However, a production commitment also requires that the structure be producible (i.e., the materials, processes, and assembly procedures are established) and the operational implications such as maintainability and repairability are fully understood and are acceptable.

No commercial airplane can enter service until initial flightworthiness certifications are issued by the Federal Aviation Administration (FAA). The FAA is also responsible for the continued safety and monitoring of the in-service fleet. Therefore, FAA participation early in the design cycle in which new material, structural concepts, and manufacturing methods are being considered will illuminate the appropriate certification and safety-of-flight issues at the point at which the most effective solutions can be applied.

The evaluations required to qualify a new material and show that manufacturing processes are robust enough to commit to production include:

- Preliminary evaluations: initial evaluation of static mechanical properties, fatigue, and corrosion resistance of the basic material; also includes assessment of the supplier's production capabilities.
- Structural performance evaluations: materials evaluation for key design issues such as fasteners and joints (e.g., static and fatigue behavior, installation damage); damage resistance and damage tolerance; corrosion (e.g., fastener installations, dissimilar materials, and finish damage); effects of shimming and sealing; component tests to verify performance; and damage caused by thermal and environmental exposures.
- Processing studies: sensitivity of key properties to expected process variations, development and validation of key manufacturing technologies (e.g., material removal, forming, bonding, finishing).
- Production readiness: fabrication of prototype parts to identify issues in fabrication and production, evaluation of aircraft repair and maintenance, and readiness of fabrication subcontractors.

The current method used for structural design and design validation for aircraft components follows a "building-block" approach. First, coupon-scale testing is performed to establish basic static and fatigue design property limits, or allowables, under pertinent environmental conditions. Second, element tests are performed to relate allowables to design elements. Finally, to verify scaling models and assumptions, tests are performed on structures from subcomponent through full-scale components, culminating in static and fatigue tests on the complete aircraft.

Test methods for developing design allowables for metallic and composite structures are fairly well established and are continually reviewed by the FAA through committees such as MIL-HDBK-5 for metals and MIL-HDBK-17 for composites so they can be refined and updated. The conditions under which the values were established, including product form, processing method, and thermal treatments, are very important. New manufacturing methods, described in chapters 3 and 4, are being developed for advanced materials. These new processing methods can produce a final material that is significantly different from previously tested materials, even though alloy designations or basic fiber and polymer chemistries are essentially the same. For example, in composites, different processes can yield significant variations in fiber and matrix volume fractions, compaction, and void content. The effect of these types of variations on materials properties must be established.

TRENDS IN MATERIALS

Advanced structural materials find increased application in new aircraft designs as the understanding of the materials and process technology matures, as the industry gains confidence in their use and production capability, and as performance needs dictate their use. The most recent example of the application of advanced materials and structures in a new aircraft is the Boeing 777, which makes use of a number of new materials, including advanced aluminum and titanium alloys and toughened polymeric composites. The 777 advanced materials applications are shown in figure II-1. For the purpose of this study, the 777 aircraft was considered to be the state of the art in new material utilization.

Airframe structural materials including metallic alloys, such as aluminum, titanium, and steel, as well as polymer-matrix composites, have evolved since their initial introductions into aircraft service (Hyatt et al., 1989). These evolutionary advances reflect improvements in both materials and processing technologies and the historical dominance of weight reduction in aircraft design.

Because of the desire to reduce overall costs (see chapter 1 discussion of cost drivers), including the costs of acquisition and maintenance, evolutionary, rather than revolutionary, material changes will be favored in the future. The principal barriers to increased use of new high-performance materials are cost of acquisition, manufacturing, certification, and product life cycle; technological risk; and the state of the materials supplier base. Aircraft manufacturers are using an integrated product development approach that considers such issues as producibility, cost, nondestructive evaluation methods and criteria, and repair and maintenance from the outset of development programs.

The committee believes that polymer-matrix composites (e.g., tailored forms; woven and sewn three-dimensional configurations; automated tape and tow placement) and advanced metallic alloys (e.g., tough aluminums; high-yield-strength aluminum; aluminum-lithium; high-strength titanium; and high-strength steel) will see increased use in next-generation commercial transport aircraft.

Innovation in structural design will take advantage of new low-cost processes, reduced part counts, and consolidated processing steps. Innovative structural concepts that enable low-cost processing methods, including integral stiffening concepts, net-shape processing (e.g., metal casting and resin transfer molding of composites), and laminated hybrid constructions are the kinds of development that will be favored for next-generation transports. Chapter 2 discusses these concepts in more detail.

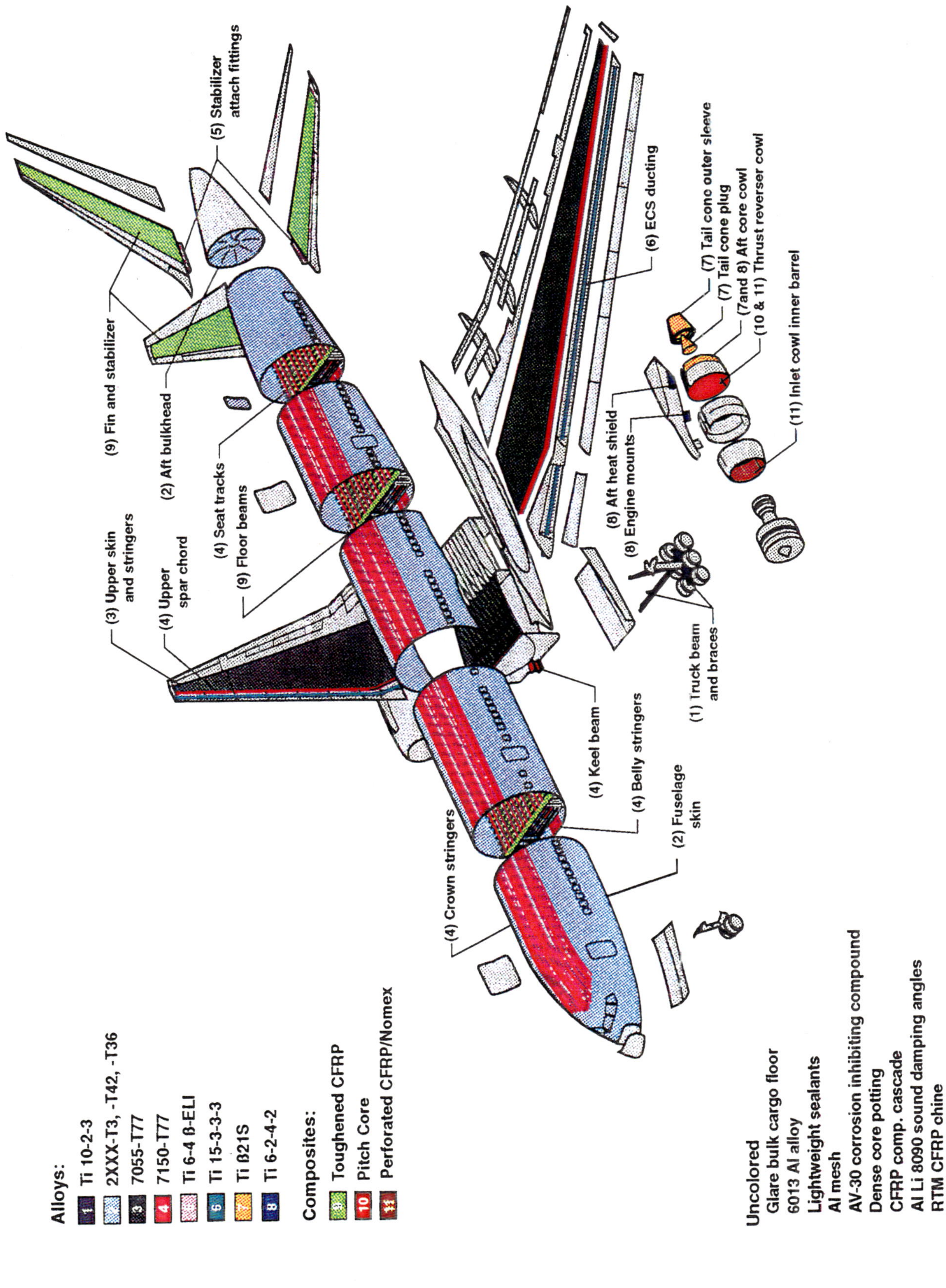

FIGURE II-1 Advanced materials on the Boeing 777. Source: Smith (1994a). Courtesy of Boeing Commercial Airplane Group.

2

Structural Concepts

The introduction of new structural concepts will depend on aircraft performance requirements, cost considerations, and the maturity of the technology. Based on these factors, as described in chapter 1, the committee believes that technology will be available for advances in next-generation applications to produce components for an advanced metallic fuselage, structural composite wing, and composite fuselage. New structural concepts will evolve as a result of requirements associated with these applications. This chapter describes design and technology issues for innovative structural concepts, as well as key design issues in the development of advanced metallic fuselage, composite wing, and composite fuselage components.

INNOVATIVE STRUCTURAL CONCEPTS

Innovative structural features are often adopted because they incorporate low-cost processing methods that take full advantage of the potential benefits possible with new materials. This section describes a number of innovative concepts and discusses the developments required to enable their application for next-generation aircraft. The potential applications, benefits, and constraints for new structural concepts including integral stiffening, net-shape components (e.g., precision castings, multiaxial construction/resin transfer molding), innovative sandwich structure, advanced joining methods, laminated hybrids, and smart materials and structures are described in the following sections and are summarized in table 2-1. Processing methods are discussed in detail in chapter 3 (metals) and chapter 4 (polymeric composites). Manifestations of such concepts in next-generation applications will be found in advanced metallic fuselage, composite wing, and composite fuselage components, as well as other applications.

Net-Shape Components

Net- or near-net-shape technology has the potential to greatly lower manufacturing costs by reducing or eliminating machining steps, by reducing the number of parts by combining subcomponents, and by producing parts in configurations that could not be easily produced using other processes. The nearest-term, most promising net- or near-net-shape processes for metals include precision forgings, net-shape extrusions, and castings. Net-shape polymeric composite processes include resin transfer molding (RTM) and resin film infusion (RFI) of multiaxial fiber preforms.

Cast structural parts can be attractive for aircraft applications due to low cost and low-weight benefits in large and complex shapes. Material-intensive operations involving machining and fastening of metal parts are minimized in the fabrication and utilization of precision castings. Efforts to reduce manufacturing costs of airframe construction in recent years have contributed to the development of economically viable processes based on advanced casting technologies, including new alloys, process controls, and simulation methods. These developments in materials and processes are described in chapter 4. Improved casting technology promises to increase the structural integrity of castings and reduce cost by eliminating the need for many small parts requiring individual manufacturing operations. Currently, the lack of confidence in cast products on the part of the structural design community has limited the use of these products in primary structural applications. Current developmental efforts in casting technologies will result in reductions in manufacturing cost, accelerate the design process for castings, provide a statistical database of critical design properties, and improve the overall quality and confidence of casting applications in primary structural components.

RTM of composite components is attractive because the process:

- reduces the need for finish machining,
- eliminates fastened or bonded joints,
- produces complex shapes in a one-step process, and
- fabricates thick parts without concern for monitoring matrix resin out-of-refrigeration time.

The application of net-shape composite processes has been hindered by poor process controls on fiber distribution and volume fraction. Development of more-reliable process controls, improvements in preform fabrication, and the availability of toughened matrix polymers that are compatible with RFI processes would enable expanded use of net-shape processes in composite primary structure.

TABLE 2-1 Summary of Innovative Structural Concepts

Structural Concept	Applications	Benefits	Constraints
Integral stiffening/ iso- and ortho-grid construction	Skin/stringer panels, doors, floor assemblies	Lower part counts, reduced use of fasteners, reduced processing steps	Part complexity, tooling costs, damage tolerance, maintenance, inspection, and repair
Precision castings	Pylons, bulkhead, stabilizers, canopy frames, doors, structural frames	Lower part counts, reduced manufacturing costs, amenable to rapid prototyping methods	Casting defects, fatigue behavior, property database
Multiaxial construction/ resin transfer molding	Skins, stringers, frames, ribs	Reduced part counts, improved delamination resistance, impact tolerance	Variable fiber/matrix distribution, property database, tooling costs
Innovative sandwich construction	Skins, control surfaces, edges, doors, floor assemblies	High strength-to-weight ratio, biaxial stability	Moisture entrapment, high manufacturing and assembly cost, low-cost in-situ inspection
Laminated hybrids	Skins, floor panels, straps	Improved fatigue resistance, stiffness	Constant thickness, inspection methods
Welded structure	Fuel tanks, pressure vessels, wing-box structure	Lower part counts, reduced manufacturing costs, amenable to automation	Low static strengths, poor fatigue behavior, property database
Adhesive-bonded structures	Wing- and empennage-box structure	Reduced fasteners, reduced manufacturing costs	Poor fatigue behavior, inadequate inspection methods

Integral Stiffening and Iso- and Ortho-Grid Construction

Integrally stiffened parts—parts in which skins and stiffeners are produced as a single unit—have the potential to significantly reduce the manufactured cost of complex components by eliminating multipart assemblies, fasteners, and adhesive-bonded joints. Integral stiffening can also reduce corrosion through the elimination of susceptible interfaces and can improve structural efficiency by providing continuous load paths and eliminating the need for joints. Machined iso- and ortho-grid construction, consisting of a skin with integral stiffening provided by intersecting machined ribs in a triangular (iso-grid) or quadrilateral (ortho-grid) pattern, represent promising approaches to integral stiffening, from the standpoint of both structural efficiency and low-cost manufacturing.

Integrally stiffened components can be produced from metallic alloys as well as polymeric composites. Metallic alloys are fabricated into integrally stiffened components using machining and chemical milling or net-shape (or near-net-shape) processes (e.g., precision forgings, extrusions, and castings). Integrally machined components are very cost-competitive because of advances in high-speed machining and improvements in cutter technologies, whereas precision forging, thick plate, and extrusions suffer from high tooling costs and difficulty in maintaining dimensional stability.

One-piece prototype iso-grid parts fabricated from several high-strength aluminum alloys have been evaluated for missile skirt and fuselage side-panel applications. Developmental efforts are underway to qualify iso-grid concepts for floor assemblies and cargo and access doors.

Integrally stiffened composite panels can be produced by co-curing (or co-bonding)[1] stiffeners with skin panels or by using net-shape processes such as RTM or RFI. Stiffened panels produced using co-curing processes are used for composite primary structure. For example, co-cured I-stiffened

[1]Co-curing is the simultaneous thermal processing of multiple uncured or partially cured details to produce one component. Co-bonding is the thermal processing of fully cured details with uncured or partially cured details to produce one component.

constructions, fabricated from carbon/epoxy, were used on the V-22 Osprey tiltrotor, on the Navy A-6 wing, and the Boeing 777 empennage. The Airbus A320 empennage has integral blade-stiffened skin panels.

The disadvantages of integrally stiffened structures are that fabrication of complex parts is difficult, tooling costs can be high, inspection (especially in the limited-access stiffener web areas) is difficult, and disassembly and repair can be very costly.

Innovative Sandwich Constructions

Sandwich constructions consist of structural facesheets bonded to a stabilizing core. The aerospace industry has developed a wide base of experience in design principles, weight and cost advantages, and field service maintainability for sandwich constructions. Although sandwich panel constructions offer unique capabilities, several potential challenges require resolution to enable technically and economically viable structural assemblies.

Sandwich structure has been proposed or used in almost every area of modern aircraft, including skins, ribs, spars, control surfaces, leading edges, doors, and floor assemblies. Most advanced aircraft have honeycomb sandwich control surfaces, and many have honeycomb access doors and panels. Many benefits are derived from the individual applications of sandwich structure. In general, sandwich panel construction can reduce weight in strength-, stability-, and stiffness-critical structures.

Sandwich panels act as very efficient I-beams. The outer facesheets are similar to flanges, accommodating both the bending loads and axial tensile and compressive loads, while the core material simulates the web by sustaining shear and compressive stresses normal to the panel. The core also prevents buckling of the facesheets under axial compressive loading. The bondline efficiently transfers the stresses from the facesheets to the core, and essentially allows the facesheets and core to act as a single structural unit. With efficient design and proper material selection, honeycomb sandwich construction often offers the minimum weight for structural configuration. Sandwich construction has also proven beneficial for thermal insulation and noise suppression.

In addition to sustaining design loads, an acceptable sandwich structure must also remain unaffected by exposure to a variety of in-service environmental and damage conditions (e.g., scratches, delamination, and impact damage). Sandwich panel construction, while offering the opportunity for minimum-weight designs through the use of minimum-gauge materials, also confers some inherent durability and damage tolerance limitations. Minimum-gauge designs with low-density cores are extremely sensitive to in-service threats such as low-velocity impacts, surface facesheet damage, and moisture entrapment. Sandwich structures typically require more maintenance because of frequent repairs in minimum-gauge designs (see chapter 7). The on-site inspection of sandwich parts (without removal from the aircraft) is usually difficult and is consequently an expensive procedure.

A wide variety of materials and products are available to the design community. The core can take the form of honeycomb, corrugated, truss, and flexible cores made from aluminum, composites, titanium, and steel as well as polymeric rigid-foam core. Preferred candidates for facesheets include aluminum, titanium, and steel alloys, as well as carbon and glass-reinforced composites and hybrids.

The use of sandwich panel construction provides a number of benefits in terms of lightweight airframe components. Several potential innovations could lead to application in critical primary structure. These include:

- Sealing materials and methods. New panel sealing concepts can reduce repair costs and potential corrosion problems associated with moisture entrapment.
- Manufacturing and assembly operations. New concepts for machining of contoured panels and core-facesheet stabilization methods may be effective in reducing the cost of producing sandwich panel constructions.
- Optimization codes for durability and damage tolerance. Improved analytical codes for durability, damage tolerance, and weight optimization with a capability for incorporating thermal and acoustical properties of sandwich constructions would provide added confidence in structural designs and could be used to develop material property goals for long-term research.
- Nondestructive evaluation procedures. Advanced nondestructive evaluation methods to ensure rapid and inexpensive evaluation of bondline integrity are mandatory for sandwich structures. Techniques that can be used on many aircraft components will help to minimize maintenance and repair costs (see chapter 8).

Advanced Joining Methods

Adhesive-Bonded Structure

Adhesive-bonded joints are used extensively in aircraft structures, especially for secondary structural applications such as core-to-skin joining of honeycomb sandwich constructions and bonding of structural details such as stringers, ribs, and reinforcing doublers. Adhesive bonding allows large-area joining and potentially reduces or eliminates the need for fastened joints. Adhesive-bonding processes are used for joining polymeric composite or metallic substrate materials. Despite the advantages, adhesive bonding has had no significant application in primary structure in commercial transports. The limited application in primary structure is because of concerns with the durability of bonded joints

(Rakestraw et al., 1994), poor nondestructive evaluation (NDE) methods (Cawley, 1992), and inconsistency and environmental impacts of surface preparation processes. The development of fracture mechanics methods to evaluate adhesive-bond durability is discussed in chapter 6.

The most prevalent adhesives used in structural bonding of aircraft structures are thermosetting film adhesives, including epoxies and nitrile-phenolics, that generally require elevated-temperature curing in a press or autoclave. Unmodified thermosets offer high strength and stiffness and moisture and creep resistance; however these adhesive systems are brittle and offer little resistance to crack growth. Thermoplastic- or rubber-modified epoxies that provide better crack growth resistance while maintaining adequate stiffness and environmental resistance have seen increased use on newer commercial transport models.

In general, the strength and durability of adhesive-bonded structure have been limited by interfacial properties, making surface preparation processes crucial. Aluminum substrates are prepared for bonding using phosphoric-acid or chromic-acid anodize processes to yield stable, porous oxide layers that aid bonding and inhibit corrosion. The anodized surfaces are coated with a corrosion-inhibiting primer to protect the surface prior to bonding, to improve the surface wetting of the adhesive, and to provide additional corrosion protection. As described in chapter 5, anodizing and priming operations are being modified or replaced to decrease their environmental impact. For instance, boric-sulfuric-acid anodize processes are replacing chromic-acid anodizing processes to eliminate the release of chromium. Water-borne primers are also being evaluated to replace solvent-based primers to reduce the release of volatile organics. The effect of these changes on bonded-joint durability and substrate corrosion resistance needs to be evaluated.

Composite substrates are prepared for bonding by mechanical abrasion (e.g., sanding or media blasting), removal of a "peel ply" (i.e., a removable reinforcing fabric co-cured onto the surface that leaves a roughened resin surface), or through electromagnetic plasma etching. Surface preparation processes for composite substrates, in general, suffer from poor consistency and reproducibility and may cause damage to surface plies. Research is continuing to develop consistent, controllable surface preparation methods for composite substrates (including composite systems with thermoset, modified thermoset, and thermoplastic matrices).

NDE methods are needed to detect:

- poor adhesion (or disbonds where substrate are in contact) between the adhesive and substrate;
- poor cohesive strength of the adhesive; and
- complete disbonds, voids, or porosity in the adhesive (Cawley, 1992).

Current methods that are most often used for commercial aircraft include ultrasonic methods and sonic vibration methods (e.g., coin tap). These methods can detect disbonds, porosity, and voids, but are difficult to perform in the field. Larger-area methods such as thermography may allow more rapid inspections. Work is needed to develop cost-effective techniques to detect poor adhesive and cohesive strength.

Welded Structure

Welding has been employed as a joining technique for aerospace structures for many years. It has been used extensively for fuel and hydraulic lines, propellant tanks, high-pressure-gas storage tanks, and missile structures. Various aluminum, titanium, and steel alloys are weldable and have been used to manufacture airframe, missile, and propulsion systems structures. With all this success, however, most airframe primary structure manufactured in the United States and Europe consist of mechanically fastened aluminum skin/stringer construction. Historically, welded structures for primary applications were not considered to be weight competitive and had poor fatigue behavior. A notable exception was the use of electron-beam welded structure on the F-14.

There are three general categories of welding processes: fusion, resistance, and solid-state. Each of these methods can be further subdivided into various processes that are commonly used in the manufacturing of aerospace structures. For most primary structural applications, electron-beam, gas-tungsten-arc, and gas-metal-arc methods have been employed, although plasma-arc and laser-beam methods typically possess advantages with respect to high energy density. Fusion welding techniques are being integrated with both robotic control heads and inspection capabilities.

Aluminum alloys that are not strengthened by heat treatment are readily weldable. However, these alloys display relatively low strengths compared with the precipitation-hardenable aluminum alloys often employed in aircraft assemblies. Welding of the high-strength aluminum alloys generally results in embrittlement, cracking, and reduced strength in the weld- and heat-affected zones. Several aluminum alloys are judged to be reasonably weldable, notably 2014 and 2219, and welding is used extensively in their primary structural applications on missile boosters and cryogenic tanks. Even in these aluminum alloys, the weldments exhibit only about 50 percent of base metal strengths, so there must be proper accommodation of strength reductions in the weld areas. Evidence of lower values of tensile elongation and fracture toughness adjacent to weldments has also led to concerns about compliance with stringent fatigue and fracture requirements. The emerging family of low-density, high-strength Al-Li alloys consists of several products that offer promising welding characteristics, particularly 8090 and 2195.

Titanium alloys are generally much more weldable than aluminum alloys. Titanium weldments can be produced with nearly 100 percent joint efficiency for strength and slight decreases in fracture and durability performance. However, titanium alloys oxidize rapidly when heated in air to temperatures in excess of 480°C (900°F). To prevent surface contamination (oxidation), titanium alloys are welded under vacuum or inert atmosphere conditions. The relatively low coefficients of thermal expansion and thermal conductivity of titanium alloys tend to minimize the potential for distortion during welding operations. All of the fusion welding processes can be used with titanium alloys, as long as the weld region is protected from extensive oxidation. Of particular recent interest is the electron-beam welding method that can join sections up to 5.1 cm (2.0 in.) thick in a single pass. Ti-6-4 is a very weldable alloy in the annealed condition, and has been employed extensively in F-14 airframe structures.

Welded structural components offer a number of potential advantages with respect to structurally efficient and affordable airframe structures. A reduction in fabrication and manufacturing costs is associated with welded structures because of lower part counts and automated assembly practices. Weight reductions are also achieved through more-efficient joints that eliminate fasteners and associated edge-margin requirements. The implementation of process controls on welded structures and the development of property databases will contribute to more-extensive utilization in future aircraft systems.

Technological developments that could expand the use of welding processes in airframe structure include:

- Optimization of welding processes. Emerging new methods such as variable polarity plasma arc, electron beam, and laser beam could lead to higher strength weldments and improved fatigue properties. High, localized heat inputs in these welding methods tend to minimize hot cracking and porosity problems.
- Automation of welding practices. The increased application of computer-controlled welding tools and inspection heads will improve process-variable control, increase the process speed and quality, and lower the cost of welded structures.
- Improved inspection technology. Inadequate methods for nondestructive inspection of complex weld joints may limit the application of welding processes.

The major pay-off for advanced welding technologies is reduced cost. Structural efficiency and lower costs will make welding more attractive than other joining processes such as mechanical fastening and adhesive bonding, although the cost of installing welding process equipment must be considered. Welding is particularly promising for fuel tanks and pressure vessels where leakproof joints are a design requirement.

Hybrid Laminate Concepts

Hybrid laminates, consisting of alternating layers of thin metallic sheet and fiber-reinforced polymeric composites, show potential for use in weight-sensitive, and fatigue- and fracture-critical components. Current variants include aluminum-based laminates using 2024 and 7075 sheet and aramid or glass fibers and titanium-based laminates using Ti-6Al-4V sheet and carbon fibers. The mechanical properties of hybrid laminate materials, like those of resin-fiber composite systems, are directional with respect to fiber orientation. Several advantages of these hybrid laminates compared with carbon/epoxy composites include plasticity, controlled residual stresses, lower material costs, and applicability of standard metal fabrication and repair practices. The outer metallic layer acts as a barrier to moisture and offers resistance to impact damage. Laminated hybrid materials exhibit outstanding fatigue properties compared with monolithic aluminum or titanium alloys. The improved fatigue performance is attributed to load transfer from the metal layers to the stronger, unbroken fibers that bridge the advancing crack and restrain further opening.

Potential aircraft applications envisioned for hybrid laminates include lower wing skins, fuselage skins, tear straps, and empennage structures. Because laminated materials display vibrational damping capabilities exceeding that of monolithic sheet, applications involving acoustic fatigue problems are encouraging. Considerable design flexibility is available with hybrid laminate materials with respect to varying stacking sequences, number of plies, and fiber orientations.

Developmental work on aluminum hybrid laminates has been performed by Delft University for more than 10 years (more recently in a joint collaborative effort with Alcoa). A number of prototype parts have been fabricated and are undergoing test evaluation by Fokker and Deutsche Aerospace. The aft cargo door of the C-17 military transport is the largest primary structural component in production fabricated from aluminum-aramid (ARALL®) laminates. Because of the success of aluminum-based laminates, considerable work is currently underway to exploit hybrid laminate benefits in titanium alloys using high-strength graphite fibers.

The family of hybrid laminates offers significant improvements over current monolithic materials for aircraft structures. For the acceptance of emerging laminate materials by the airframe industry, the following issues must be addressed:

- Design flexibility. Analysis and manufacturing experience with laminate materials must be developed to accommodate complex shape and thickness changes typically encountered in airframe components.
- Manufacturing cost. Methods to reduce or off-set the higher material costs of hybrid laminates require development and demonstration in the production environment. Ply drop-off and sheet-splicing techniques are

likely to be important for large fuselage and wing applications of these laminate materials.

- NDE. Low-cost, in-situ inspection methods are necessary to locate and quantify damage incurred under service conditions. Repair and maintenance procedures must be developed and proven for large-scale laminate components.

Smart Materials and Structures

Smart, or adaptive, materials and structures are a recent concept that is still rapidly evolving and offers significant benefits in a range of applications including aircraft design and performance. The development of smart materials is being enhanced by significant support from the Advanced Research Projects Agency of the U.S. Department of Defense. In the recent National Research Council report, *Expanding the Vision of Sensor Materials* (NRC, 1995b), a considerable discussion is devoted to the definition and application of smart materials to structural monitoring. The report cites a classification of smart materials developed by Newnham (1993). Under this classification, smart materials range from:

- passive smart materials that respond to external change without external control;
- active smart materials that utilize a feedback loop to enable them to function like a cognitive response through an actuator circuit;
- very smart materials that sense a change in the environment and respond (e.g., by altering one or more of their property coefficients, tuning their sensing, or actuation capabilities); and
- intelligent materials that integrate the sensing and actuation functions with the control system.

In the ultimate manifestation, smart structures will be composed of an array of sensors, actuators, and signal processors, some of which will be the smart materials themselves, that are linked to cause the structure to perform in a predetermined fashion to external stimuli.

Smart materials and structures can be thought of as mimicking the human biological system. For example, just as the brain, through the nervous system, can command the muscles to take some action, it is conceivable that a flight control system (sensor and processor) could send an electrical signal to cause an actuator to deflect a fin or change the aerodynamic shape of a wing. Potential applications of smart materials and structures relevant to advanced aircraft include (Crowe, 1994):

- control of aerodynamic surfaces (e.g., adaptive wing);
- damping and tuning of structures (e.g., rotorcraft blade damping);
- noise control, both radiated engine noise and internal cockpit and cabin noise;
- self-diagnostic capability;
- robotics;
- embedded avionics; and
- de-icing of aerodynamic surfaces.

The elements of a fully integrated smart materials system include sensors, signal processors, actuators, and its own power supply. Currently, a range of potential actuator concepts is being developed, including piezoelectric, electrostrictive, and magnetostrictive actuation, as well as shape memory materials. Shape memory materials, based on Ni-Ti alloys, operate over a limited frequency range (up to 1–5 Hz) but have considerable actuation displacement (up to ~8 percent strain), whereas, in contrast, the electro-ceramic actuation concepts have limited displacement (~0.1 percent) but offer essentially unlimited frequency response up to the megahertz range.

Key issues in the development of a smart materials system for commercial aviation applications still relate to cost and reliability of actuators, system weight, and overall complexity. In the present harsh economic climate, these smart materials systems will not be adopted unless they significantly enhance aircraft performance or improve safety or reliability. However, the eventual possibility of significant benefit seems real. One example presented to the committee is that the addition of adaptive control surfaces to a military plane could allow the elimination of hydraulics and hinge lines, providing significant advantages in increased payload, range, and maneuverability (Crowe, 1994).

In another example of potential benefit, the application of smart materials to suppress blade vibration and control twist in rotorcraft could enhance speed capabilities and reduce required maintenance (Crowe, 1994).

Another possible application, which may see early implementation in commercial aviation, involves vibration cancellation to reduce cockpit noise. This application, while it is difficult to quantify its benefit, would potentially improve safety by reducing the fatigue of pilots.

The committee does not believe that, in the approximately 20-year purview of this report, systems involving adaptive materials will be available, practical, and sufficiently low enough in cost to enable application on commercial aircraft. However, the sensing and signal-processing capabilities could be available for health monitoring of structures or measurement of service environmental conditions. For example, it is considered likely that smart materials will first see usage in the most simple form, that is, as a passive system. For instance, optical fibers could be used as an integral part of a health monitoring system to assess the performance of critical components in locations that are difficult to inspect. Optical fibers can be used extrinsically to merely relay information from a specific sensor (a corrosion sensor connected to the end of the fiber). Alternatively, the optical fiber could be applied intrinsically, where changes of the light-transmitting properties of the fiber itself (which may be embedded in a component) can be used to provide information of the amount of strain imparted to the component. Such a distributed system of sensors and optical fibers could then provide a rudimentary health monitoring approach. Technical issues associated with such a passive system would involve the selection of key components for monitoring, the embedment or attachment of the fiber within or on a structure to minimize fiber breakage, and some means of calibrating the fiber and its responses in order to correctly interpret the information obtained.

To summarize, the field of smart materials is now being extensively researched and the technology is developing rapidly. Accordingly, the committee recommends that the FAA closely monitor developments in this technology and attempt to quantify the performance enhancements that may be possible through the application of smart materials systems. The fabrication of a prototype or demonstration unit would be one means of thoroughly evaluating the performance and benefits of a smart materials system.

ADVANCED METALLIC FUSELAGE

A logical step in the technological evolution of commercial aircraft is the incorporation of advanced structural alloys and innovative manufacturing processes in fuselage applications. Design, fabrication, and testing of an advanced metallic fuselage structure is needed to validate reduced manufacturing costs and improved structural efficiency for new candidate materials and innovative design concepts. In general, cost savings derive from reducing labor-intensive processes and lowering part count for major assemblies.

The structural performance and associated manufacturing cost baselines for several new design concepts require quantification and validation prior to acceptance. Candidate materials systems include new high-strength, high-toughness, and low-density aluminum- and titanium-based alloy systems discussed in chapter 3. Hybrid laminate composite materials also offer very attractive, high-specific properties with the attendant benefit of significantly reduced crack growth rates (Bucci et al., 1989; Gregory and Roebrocks, 1991).

Innovative processing approaches to forming, joining, and finishing systems may offer cost savings. Innovative processing methods for advanced metallic materials are described in chapter 3. These new processing approaches should be evaluated in comparison with conventional skin/stringer construction costs and structural efficiencies. Design and manufacturing issues to be addressed are described in table 2-2.

Testing and demonstration components should evaluate a number of new design concepts such as adhesive-bonded components, hybrid laminate composite materials, superplastically formed and superplastically formed diffusion-bonded components, and welded assemblies. The utilization of hybrid laminates in fuselage structures will depend greatly on optimization of ply orientations, drop-offs, and improved mechanical property behavior. The building-block approach described in the introduction to part 2 will be used to fabricate and test specimens at incremental levels from simple coupons to components to full-scale structural validation components.

TABLE 2-2 Design and Manufacturing Issues in Advanced Metallic Fuselage Development

Design
Design for manufacture and assembly principles
Effects of modulus mismatch with adjacent structures
Reduction in number of parts
Low-cost joint designs
Test and Analysis
Durability and damage tolerance of integral structures, laminate materials, and bonding techniques
Verification of analysis methods
Manufacturing
Tooling and fabrication concepts and processes
Nonautoclave bonding methods
NDE methods for new fabrication concepts

COMPOSITE WING

Based on service experience from composite stabilizer components, the next most likely application of composites in primary structure is the wing. The experience gained in designing and operating composite horizontal stabilizer components for the Boeing 737 NASA flight-service evaluation, the Airbus A320, and especially the Boeing 777 is valuable in the development of composite wing structure, since these components are structurally similar. However, commercial aircraft wing structures have additional design considerations, including fuel containment, electrical grounding, engine mounting, and fuselage "carry-through" structures. The critical design and manufacturing issues are summarized in table 2-3. Composite wing structures have been investigated as part of NASA's Advanced Composite Technology (ACT) program, led primarily by a team from McDonnell Douglas.

Composite wing structure is particularly attractive because the high strength and stiffness and low density of composites offer the opportunity to produce long, slender (high aspect ratio) wings that would not be producible using monolithic materials.

Several approaches could be used to produce a composite wing structure. One approach would base the design on the technology used for the Boeing 777—toughened skin/stringer with ribs—to take advantage of the design

TABLE 2-3 Design and Manufacturing Issues in Composite Wing Development

Design
Wing-to-body intersection
Engine attachments
Fuel containment and electrical grounding
Metal-to-composite interfaces
Damage tolerance of upper skin panel
Manufacturing
Scaling of low-cost processes
Technology development of integral and co-cured stringers
Inspection and repair technologies

simplifications afforded by the toughened system. Improved processing technologies (e.g., automated fiber placement for skin lamination and resin transfer molded stiffeners) would be needed to make the application cost-competitive.

Another manufacturing approach is being investigated by McDonnell Douglas under the ACT program to produce a near-net-shape, integrally stiffened panel using RFI processing (Janicki, 1994). The structure would be fabricated from stitched fiber preforms and infiltrated with low-cost resin-matrix systems similar to those used on previous-generation secondary structures. Damage tolerance and delamination resistance would be imparted to the structure via the through-thickness reinforcement provided by stitched fibers. Scaled components have been produced using this technique. Work is continuing on high-speed sewing equipment and RFI (resin film intrusion) process optimization.

COMPOSITE FUSELAGE

There has been a significant research effort directed toward the design and validation of a polymeric composite fuselage (Smith et al., 1994). Factors influencing manufacturing costs for a composite fuselage are shown in figure 2-1. Although many cost drivers affect the entire fuselage section, each area of the fuselage (crown, side, and keel) has unique challenges that must be overcome to achieve low costs.

Important composite structural design issues are also unique to each of the fuselage areas as shown in figure 2-2. Design of the crown panel is primarily governed by tension-loading requirements. In contrast, the side area is dominated by shear and pressure-load redistribution around the door and window cut-outs. Design of the keel is driven by complex load redistribution from the keel beam and combined loads dominated by axial compression.

Emerging manufacturing technologies for composite fuselage structures are presented in figure 2-3. As can be seen in the figure, development emphasis is being placed on automated techniques such as advanced tow placement for skin fabrication, textile preform and RTM (resin transfer molding) technology for the substructure, and co-curing and co-bonding operations for assembly. The majority of the research in this area is being conducted by the Boeing Commercial Airplane Group under the Advanced Technology Composite Aircraft Structure (ATCAS) program as part of NASA's ACT initiative.

Ten items have been identified by the ATCAS program as representing critical issues that must be addressed to make composites a viable design option for fuselage structure

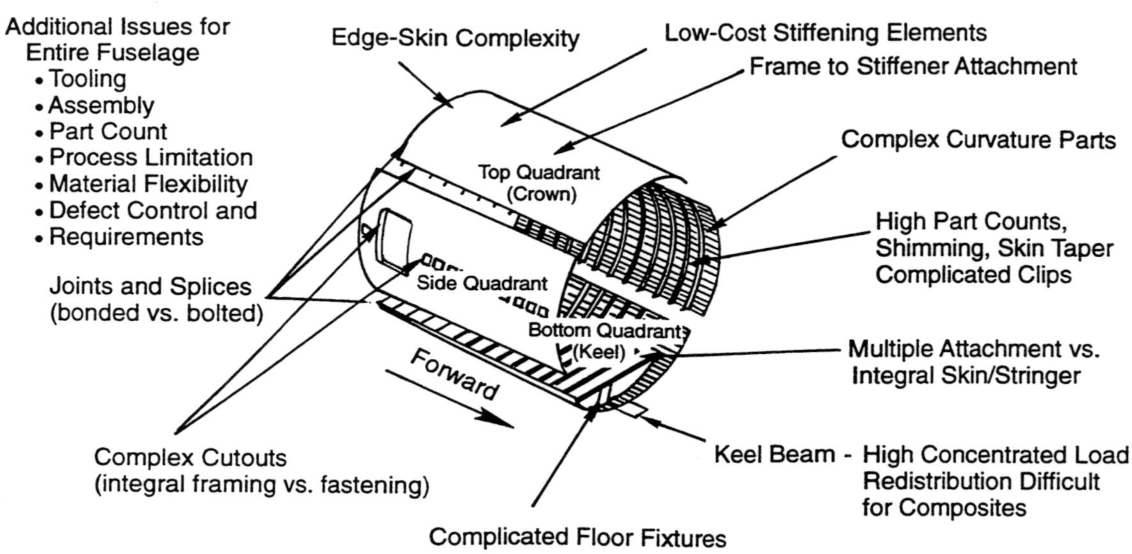

FIGURE 2-1 Cost drivers for a composite fuselage. Source: Ilcewicz et al. (1990).

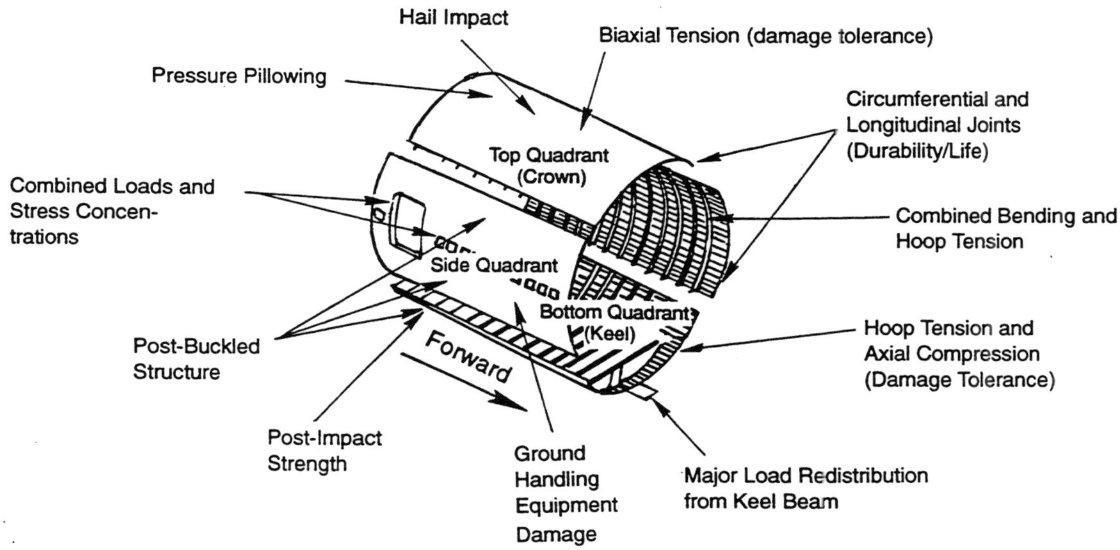

FIGURE 2-2 Structural design drivers for a composite fuselage. Source: Ilcewicz et al. (1990).

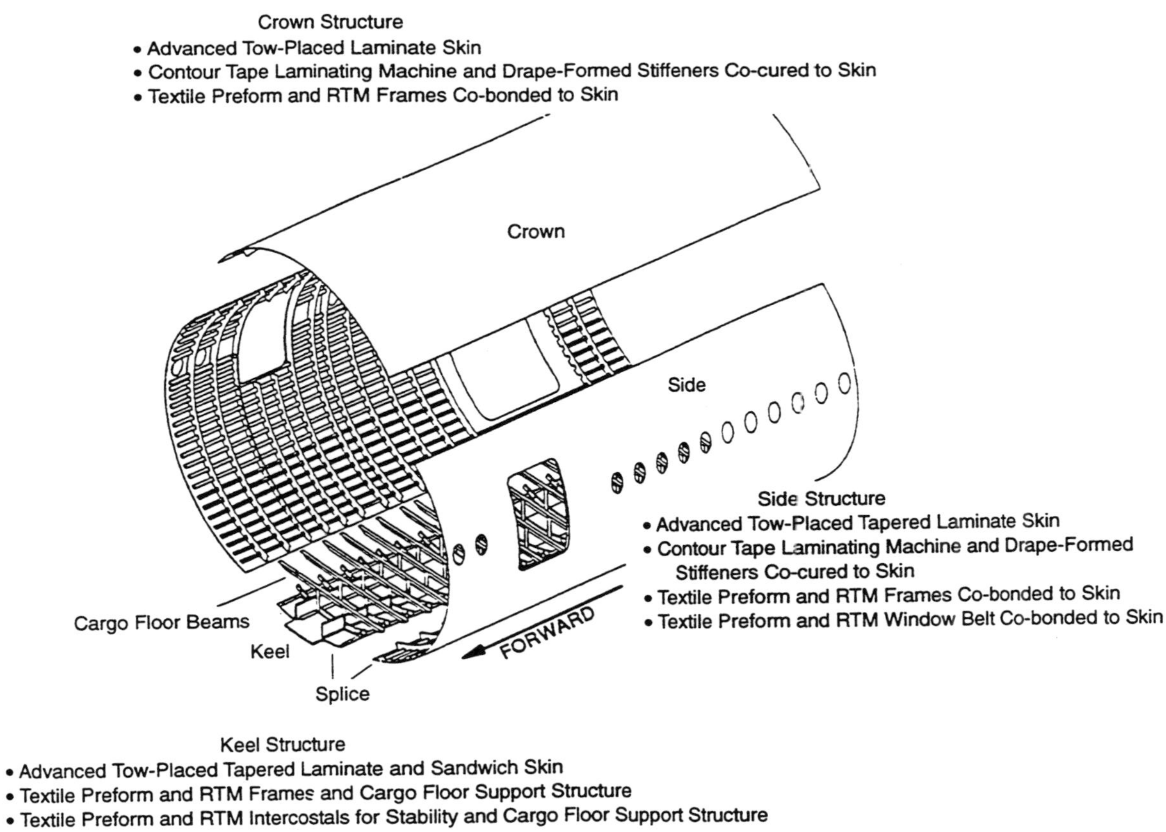

FIGURE 2-3 Emerging manufacturing technologies for composite fuselage structure. Source: Flynn et al. (1993).

(Ilcewicz et al., 1992). These are summarized in table 2-4. A majority of these issues are currently being addressed by the airframe industry under NASA's ACT program. Detailed discussions of these technical issues can be found in Ilcewicz et al. (1992). The industry goal of demonstrating technology readiness will be achieved when these major technical issues have been resolved in sufficient detail to provide the necessary confidence for the commitment of composites to commercial transport fuselage applications.

TABLE 2-4 Design and Manufacturing Issues in Composite Fuselage Development

Design
Damage tolerance of crown, keel, and side panels
Load redistribution near major fuselage cut-outs
Wing-to-body intersection
Integrity of bonded elements in configured fuselage structures
Development of mechanical joints for major panel splices
Metal-to-composites interfaces
Manufacturing
Manufacturing scale-up of configured panels
Inspection and repair technologies for selected designs
Technology developments for low-cost framing elements
Structural detail and manufacturing cost relationships for selected designs and processes

SUMMARY

Innovative structural concepts will be developed for next-generation commercial transport aircraft to take advantage of improved cost and performance benefits offered by the application of new materials that enable low-cost processing methods. This chapter describes a number of innovative concepts and discusses the developments required to enable their application on next-generation aircraft. Manifestations of advanced structural concepts will be found in components for the advanced metallic fuselage, composite wing, and composite fuselage structures, as well as other applications. The introduction of new structural concepts will depend on aircraft performance requirements, cost considerations, and the maturity of the technology. Promising structural concepts include net-shape technology, integral stiffening, innovative sandwich structure, advanced joining methods, laminated hybrids, and smart materials and structures.

Net- (or near-net-) shape technology (including precision forgings, net-shape extrusions, and castings for metals and resin transfer molding and resin film infusion of polymeric composites) has the potential to lower manufacturing costs by reducing or eliminating machining, reducing the number of parts, and by producing parts that cannot be produced using other methods.

Integrally stiffened parts—parts in which skins and stiffeners are produced as a single unit—have the potential to significantly reduce the manufactured cost of complex components by eliminating multipart assemblies, fasteners, and adhesively bonded joints. Integrally stiffened components can be produced from metallic alloys (using machining and chemical milling or net-shape processes), as well as polymeric composites (using co-curing, co-bonding, or net-shape processes).

Sandwich constructions—structural facesheets adhesive-bonded to a stabilizing core—generally provide a minimum-weight solution for strength, stability, and stiffness-critical structures. Minimum-gauge designs with low-density cores are extremely sensitive to in-service threats such as low-velocity impacts, surface facesheet damage, and moisture entrapment. Sandwich structures typically require more maintenance due to frequent repairs in minimum-gauge designs and difficult and expensive on-site inspection techniques.

Adhesive-bonding allows large-area joining and potentially reduces or eliminates the need for fastened joints. Adhesive-bonding processes are used for joining polymeric composite or metallic substrate materials. Despite the advantages, adhesive bonding has had no significant application in primary structure in commercial transports. The limited application in primary structure is because of concerns with durability of bonded joints, poor nondestructive evaluation methods, and inconsistency and environmental impacts of surface preparation processes.

Welded structural components offer a number of potential advantages with respect to structurally efficient and affordable airframe structures. Various aluminum, titanium, and steel alloys are weldable and have been used to manufacture airframe, missile, and propulsion systems structures. A reduction in fabrication and manufacturing costs is associated with welded structures through lower part counts and automated assembly practices.

Hybrid laminates show potential for use in weight-sensitive and fatigue- and fracture-critical components. Hybrid laminates offer fatigue crack growth resistance that is significantly better than monolithic aluminum and titanium alloys. In addition, hybrid laminates can be produced using standard metal fabrication processes and have improved resistance to absorbed moisture and impact damage compared with polymer-matrix composites. Potential aircraft applications envisioned for hybrid laminates include lower wing skins, fuselage skins, tear straps, empennage structures and applications involving acoustic fatigue.

Smart materials and structures, which incorporate actuators connected to both sensors and signal processing to respond to some external stimuli and cause some action or control function to occur, are a recent concept that is still rapidly evolving and offers significant benefits in a range of applications including aircraft design and performance. The committee does not believe that systems involving adaptive materials will be available, practical, and sufficiently low in cost to enable application on next-generation commercial aircraft. However, the sensing- and signal-processing capabilities could be available for health monitoring of structures or measurement of service environmental conditions.

3

Metallic Materials and Processes

Aircraft alloy materials and processing technology has been advancing steadily with each new aircraft model. Important alloys in commercial transport applications include high-performance aluminum alloys, high-strength steels, and titanium alloys. Significant progress is being made in developing alloys with improved strength, toughness, corrosion resistance, and producibility. Advances have been achieved primarily through incremental improvements to already-developed alloys. Manufacturing process development has been emphasizing low-cost approaches such as net-shape processing (casting and forging), improved forming methods, and high-speed machining. This chapter describes developments and trends in metallic alloy materials and processes.

ALUMINUM ALLOYS

The primary use of high-strength aluminum alloys is in aircraft construction; the airframe of modern aircraft is approximately 80 percent aluminum by weight (Marceau, 1994). Traditionally, the structural aluminum alloys in aircraft have been 2024 in damage-critical areas and 7075 in strength-critical areas (Starke and Staley, 1996).

The goal of aircraft designers to improve durability and save weight has led to the development of new aluminum alloys that provide improved combinations of specific strength, durability, and damage tolerance. Most often, the newer alloys are variants of older 2XXX- and 7XXX-series alloys—but with tighter controls on chemistry and processing parameters. For instance, the upper and lower wing structures of the Boeing 757 and 767 are manufactured with improved alloys relative to the older Boeing 747. The improved alloys include 7150-T6 plate and extrusions (upper wing) and 2324 plate and 2224 extrusions (lower wing). Alloy 7150-T6 is a modification of the 7050-T74 product. It is aged to a higher strength and is processed to control the grain structure and degree of recrystallization (Staley, 1992). The T6 temper results in a higher strength than the T74 temper, and a new aging treatment (T61 temper) was developed to provide one letter-grade better in the rating system used to describe exfoliation corrosion. Alloy 7150-T61 plate and extrusions are also used on the McDonnell Douglas MD-11. The tighter controls on chemistry and processing parameters may cause an increase in the cost of the material, but production applications of improved alloys show that this cost can be offset by benefits in performance or durability.

Improved Strength and Corrosion Resistance

The conventional T76 and T73 tempers used to develop high resistance to exfoliation corrosion and to stress corrosion cracking in the short transverse direction of 7XXX alloys are associated with a 10–15 percent reduction in strength compared with the peak-aged T6 temper. To address this reduction, the T77 temper has been developed for the 7150 and 7055 alloys. The 7150-T77 plate and extrusions have the strength and fracture toughness of 7150-T6 and -T61, but with the exfoliation and stress corrosion resistance of 7075-T76. The alloy 7055 relies on strict control of solute elements and thermomechanical processing to produce a material that has a higher strength than that of 7178-T6, along with improvements in exfoliation corrosion, stress corrosion cracking susceptibility, fracture toughness, and fatigue resistance (Staley, 1994).

Improved Durability and Damage Tolerance

The alloys 2324-T39 and 2224-T3 were developed by modifying the composition and processing of standard 2024. The amount of cold work applied after quenching and prior to aging was increased from the 1–3 percent used for 2024-T351 plate to about 9 percent. The allowable limits of iron and silicon impurities were reduced, and composition and processing were modified to minimize constituent particles and to improve fracture toughness and decrease fatigue crack growth rate. Processing conditions were also modified for extrusions in order to retain the deformation crystallographic texture for added texture strengthening (Staley and Rolf, 1993).

Boeing is using a new Alcoa alloy, C188, for the fuselage of the 777. This alloy falls into the 2XXX-series family and has stricter chemistry and process controls than normal airframe alloys. It has a 17 percent improvement in toughness and a 60 percent slower fatigue crack growth compared with 2024-T3.

Since the aluminum alloys discussed above are variants of or improvements over conventional aluminum alloys that are currently used—some have been used for over 50 years (2024-T3 sheet was used as early as 1936 on the original DC-3)—they minimize potential risks for design, maintenance, and in-service failure. On the other hand, new classes of aluminum alloys (e.g., Al-Li alloys and aluminum/polymeric composite hybrids described in a later section) are quite different from conventional aluminum alloys and must be treated differently.

Low Density (Aluminum-Lithium) Alloys

The interest in Al-Li alloys derives from the large effect that lithium additions have on the modulus of aluminum, a 6 percent increase for every weight percent added, and the density, a 3 percent decrease for every weight percent added. These changes apply for lithium additions up to 3 weight percent. There have been three generations of Al-Li alloys (Starke and Blankenship, 1993):

- Those produced in the 1950s, 1960s, and 1970s, including alloy 2020 (used for the upper and lower wing skins of the North American RA-5C Vigilante aircraft) and alloy 1420 (used for a welded fuselage and cockpit of the MIG 29). These alloys experienced either ductility and fracture toughness problems (2020) or were of relatively low strength (1420).
- Those produced in the 1980s, including alloy 2090, 2091, 8090, and 8091. These alloys had an attractively high modulus and low density, but contained anisotropic mechanical properties.
- The more recent high-strength Weldalite®-type alloys developed by Martin Marietta, including alloy 2195 (Pickens et al., 1991).

Aluminum-lithium alloys have yet to replace conventional aluminum alloys in many aerospace applications because of their fracture behavior. Low fracture toughness was a concern with the first commercial Al-Li-X alloy 2020. Although not all Al-Li-X alloys suffer from low toughness, fracture behavior can depend on a number of variables (Blankenship and Starke, 1992), including:

- the presence of tramp elements such as sodium and potassium (and in some cases calcium, hydrogen, and sulfur) that are believed to be present in lithium metal used in preparation of the alloy;
- the presence of iron-, silicon-, copper-, and magnesium-rich constituent phases that form during the casting process;
- the effect of chromium, manganese, and zirconium dispersoids which are added for grain structure control;
- anisotropic behavior resulting from lamellar grain structure and intense deformation texture;
- strain localization from the cutting of shearable matrix precipitates (primarily Al_3Li); and
- strain localization from preferential deformation in softer, precipitate-free zones (PFZ) adjacent to grain boundaries.

As a result, newer Al-Li alloys have been developed that minimize these effects and improve fracture toughness. Tramp element levels have been decreased by using high-purity lithium metal for the production of commercial ingots and by using fluxing and other specialized casting processes. Although the presence of constituent phase particles in conventional and Al-Li alloys is unavoidable, toughening strategies revolve around minimizing their volume fraction by keeping iron and silicon levels to a minimum. A new alloy, designated AF(UDRI), which is produced using special alloying and processing procedures that reduce anisotropy, has been developed at the University of Dayton Research Institute. The new alloy has mechanical properties similar to 7050-T7451 with significant improvements in density and modulus. Most of the newer Al-Li alloys have alloying elements (e.g., copper, magnesium, and manganese) that reduce shear localization in the matrix by forming nonshearable precipitates. Finally, strain localization in the PFZ and associated void nucleation at grain boundary precipitates cannot be eliminated by alloying additions. However, this problem can be minimized by stretching the material and then aging the material at low temperatures, both of which accelerate precipitation in the matrix and lead to decreased grain boundary precipitation, a smaller PFZ, and improve combined strength and fracture toughness.

The first and second generation Al-Li alloys (e.g., 2020, 2090, 2091, and 8090) suffered from low fracture toughness in the short transverse direction and anisotropic properties. However, newer alloys with lower lithium contents (e.g., Weldalite) and alloys that utilize special processing procedures (e.g., AF(UDRI)) have fracture toughness similar to that of many conventional aerospace aluminum alloys and offer lower densities and higher modulus.

The newer classes (second and third generations) of Al-Li alloys have seen limited commercial use. Alloy 8090 plate, extrusions, and forgings were used in at least one classified U.S. space vehicle and currently are used on leading edges and outer lower wing skins of the Airbus A330 and A340. Alloy 2090 is being used on the C-17 military transport. The Weldalite 2195 alloy has been selected for the super-lightweight external fuel tank of the Space Shuttle. This alloy offers a 50 percent increase in strength, a 5 percent increase in elastic modulus, and a 5 percent reduction in density compared with the conventional 2219 alloy that it will replace. It is likely that this alloy will also see use in advanced subsonic aircraft. Al-Li alloys may behave differently from conventional aluminum alloys, and, accordingly, they should be characterized to gain confidence in manufacturing methods and performance over the life of an aircraft.

The Al-Li alloys as a group have attractive fatigue properties; are amenable to superplastic forming; display moderate to good weldability; and can be chemically milled, bonded, anodized, clad, and painted. On the debit side, they often display considerable anisotropy, especially in the short transverse direction, and are typically more costly than conventional aluminum alloys. They are more susceptible to surface oxidation and are prone to warping during quenching. They are more difficult to process than conventional aluminum alloys, and their properties can be greatly influenced by relatively minor processing variations.

New Materials

There are a number of new aluminum materials that are under study and are being developed for use in commercial transport aircraft. Laminated hybrids of aluminum sheet with aramid-fiber-reinforced (ARALL) or glass-fiber-reinforced (GLARE®) composites have high fatigue resistance and the potential for significant weight savings in aircraft. This material also has resistance to burnthrough in the event of a fire and can potentially substitute for titanium in fire walls. On the negative side are the very high material costs, typically 7–10 times that of monolithic aluminum sheet (Tenney, 1992).

Al-Mg-Sc alloys, while potentially very expensive because of the presence of scandium, appear to have excellent corrosion resistance, and as a body skin they may not need to be clad or painted which would lead to reduced maintenance costs. Higher-strength forgings, age-formable alloys, less-quench-sensitive alloys, and rivet alloys with improved formability are all being examined and developed for future use in subsonic aircraft. Each product may require new methods of evaluation and maintenance. The critical properties that characterize the materials, fabrication and assembly, and issues of in-service supportability must be identified and evaluated prior to airframe application.

HIGH-STRENGTH STEELS

Research on the development of new high-strength, high-toughness, corrosion-resistant steels for landing gear materials has been a subject of intense recent interest. Improved Ni-Co, low-carbon steels (most notably Aermet 100 and AF1410), have excellent combinations of properties and are developed to the point where they are now being specified as replacements for the standard landing gear steels 300M and 4340.

These improved steels are used in landing gear on carrier-based aircraft because they exhibit excellent damage tolerance and environmental resistance. The steels can also find application as attach fittings, horizontal stabilizer spindles, arresting-hook shanks, and catapult hooks. Other aerospace applications under consideration include rotorcraft actuators and masts, gas turbine engine shafts, and rocket motor casings. Nonaerospace applications include ordnance, armor, high-strength fasteners, pump splines, and automotive drive shafts.

The improved combination of strength, damage tolerance, and stress corrosion cracking resistance provides significant benefits for applications under severe service conditions, such as the naval aircraft environment. An additional benefit includes fatigue strength superior to 300M. These materials have good weldability because of low carbon content. Testing has shown that Aermet 100 can be welded, without preheat, with joint efficiencies approaching 100 percent.

The new landing gear steels are more resistant to stress corrosion cracking than 300M, but are prone to general corrosion attack. Coating technology to prevent general corrosion in these steels is lagging. More work should be done in chemical vapor deposition and physical vapor deposition techniques for applying coatings. These improved toughness steels require vacuum induction melting and vacuum arc remelting practices, followed by thermomechanical processing of the wrought materials to produce the desired fine-grain size and combination of properties. Furthermore, components must be processed oversized to avoid decarburization, and processing and melting practices must be standardized. Appropriate weld filler metals are available, and full characterization of weldability of these alloys will increase their application potential.

It appears that the strength of these new steels cannot increase to higher levels without corresponding decreases in ductility and toughness. However, increased strength can be achieved while keeping toughness at levels acceptable for many applications. Such a balance of properties may be acceptable for landing gear for civil aircraft.

The use of rapid solidification technology may provide an avenue for further improvements in landing gear steels by decreasing inclusion size. However, large forging presses would be required to consolidate billets large enough for landing gear components. Powder-particle oxide coatings must be broken up during consolidation to minimize the size of oxide particles present in the finished material and to limit their effect on the mechanical properties. The need to employ thermomechanical processing may limit applications. In some cases, the desired component size may exceed the size of available furnace capacity.

TITANIUM ALLOYS

Titanium and titanium alloys are widely used in aircraft applications because of their high strength-to-weight ratio and excellent corrosion resistance. Titanium use is, however, strongly limited by its higher cost relative to competing materials, namely aluminum alloys and steels. There are three types of titanium alloy systems based on the composition of the alloy and the resultant, predominant room-temperature phase: (1) α and near-α alloys, (2) α/β alloys, and (3) β alloys. α phase is the low-temperature allotrope of titanium and phase is the high-temperature allotrope. Manipulation of the content and microstructural form of these two phases through alloying and thermomechanical processing is the primary basis for the titanium alloy optimization. For a more detailed discussion of basic titanium metallurgy, see Collings (1984), Duerig and Williams (1984), and Bania (1993).

The primary reasons for using titanium for aircraft applications include:

- Weight savings. In various applications (assuming they are not gauge-limited) the strength/weight ratio of titanium can exceed those of stronger but heavier steel alloys and lighter but weaker aluminum alloys. The consequent weight savings achieved by using titanium instead of the competing alloys can be significant.
- Operating temperature. Titanium is most commonly used when the operating temperature exceeds about 135°C (275°F; the normal maximum operating temperature for aluminum). These conditions exist in the nacelle, auxiliary power unit area, and wing anti-icing systems for airframe structures. Steel and nickel-base alloys are obvious alternatives, but have a density of about 1.7 times that of titanium.
- Space limitation. Titanium may replace more-easily processed aluminum alloys where space is limited (e.g., landing gear beams).
- Corrosion resistance. Excellent corrosion resistance enables titanium to be used, in most applications, without the addition of protective coatings.
- Composite compatibility. Titanium has found significant use in contact with polymeric composite components because titanium is more galvanically compatible with carbon fibers than aluminum and has a relatively good match of thermal expansion coefficients.

α and Near-α Alloys

There are two types of alloys in this category—the commercially pure grades (with oxygen and iron as the primary alloying elements) and those with intentional additions of α stabilizers, such as aluminum and tin. Commercially pure (CP) grades can be obtained with minimum yield strengths from 25–70 ksi (172–482 MPa), with the higher-strength grades containing more oxygen and iron. Their primary attributes are good formability, with the formability decreasing as the strength increases; excellent corrosion resistance; and good weldability. CP alloys are used for nonstructural applications such as floor support structure in the galley and lavatory areas, tubes or pipes in the lavatory system, clips and brackets, and ducting for the anti-icing and environmental control systems. CP alloys will continue to be used in commercial aircraft, with little change anticipated for future aircraft.

The α and near-α structural alloys include Ti-3Al-2.5V (Ti-3-2.5), Ti-5Al-2.5Sn (Ti-5-2.5), Ti-8Al-1Mo-1V (Ti-8-1-1), and Ti-6Al-2Sn-4Zr-2Mo (Ti-6-2-4-2S). Ti-3-2.5 has two key applications in aircraft: hydraulic tubing where it is used for high-pressure hydraulic lines, and honeycomb core in applications where greater strength than CP alloys is required.

The most common application of α alloys (other than CP) is for elevated-temperature applications because of their outstanding elevated property retention and creep resistance. The primary alloy used in the United States for these applications is Ti-6-2-4-2S, which is used for rotating engine components such as blades, discs, and rotors at temperatures up to about 540°C (1000°F). Ti-6-2-4-2S is also finding applications in some airframes, in areas such as engine mounts, exhaust systems, and areas of exhaust impingement.

Future trends in alloys are use in improved oxidation resistance and high-temperature creep strength. For example, a modification of Ti-6-2-4-2S, Ti-6Al-2.8Sn-4Zr-.4Mo-.4Si (Timet Timetal-1100®), has the potential to be used at temperatures up to 593°C (1100°F; Bania, 1989).

α/β Alloys

α/β alloys, including Ti-6Al-4V (Ti-6-4), Ti-6Al-6V-2Sn (Ti-6-6-2), and Ti-6Al-2Sn-4Zr-6Mo (Ti-6-2-4-6), are capable of somewhat higher strengths than the near-α alloys, have good combinations of properties, are more processable than the alloys, and, depending on the alloy, are sufficient for service temperatures up to 315–400°C (600–750°F). Alloys with lower β-stabilizing contents, such as Ti-6-4, are highly weldable. As the β-stabilizer content increases or as the hardenability increases, welding becomes more difficult. These alloys can provide a weight savings with superior corrosion resistance compared with low alloy steels and aluminum alloys.

Ti-6-4 is the workhorse of the titanium industry; it accounts for about 60 percent of all titanium production and 80–90 percent of the titanium used in all sections of the airframe (including fuselage, nacelles, landing gear, wing, and empennage). Virtually all product forms are used, including forgings, bar, castings, sheet, plate, extrusions, tubing, and fasteners.

Alloy Ti-6-6-2 was used extensively in the landing gear support structure of the Boeing 747 because of its superior corrosion resistance to the low-alloy steels. One β alloy, Ti-10V-2Fe-3Al (Ti-10-2-3), has been used on later models because it offers improved weight savings, particularly in thick sections. Fracture toughness and stress corrosion resistance are also improved beyond that of Ti-6-6-2.

Future trends in α/β alloys include the introduction of alloys (e.g., Ti-6Al-2Sn-2Mo-2Zr-2Cr) with improved strength/toughness combinations and the use of lower oxygen content versions of Ti-6-4 for maximum toughness.

β Alloys

β alloys, which include Ti-15Mo-2.7Nb-3Al-0.2Si and Ti-3Al-8V-6Cr-4Mo-4Zr (β-C), are capable of being heat treated to high strengths in excess of 200 ksi (1,378 MPa). They can be heat treated over a broad range of strengths, permitting one to tailor the combination of strength and fracture toughness properties that is desired, and they generally have high stress corrosion resistance. β alloys offer fabrication advantages, particularly for producing sheet, due to their cold-rolling capabilities. In addition, for hot-die or isothermal precision forgings, alloys such as Ti-10-2-3 can be forged at lower temperatures, resulting in lower die costs and forging advantages for some shapes.

Some of the β alloys, such as Ti-10-2-3 (Boyer, 1980, 1993, 1994; Carey et al., 1985; Davies, 1993) and β-C (Boyer et al., 1984; Eylon et al., 1988; Wagner and Gregory, 1993) have excellent fatigue properties, while others, such as Ti-15-3, have, in general, poor fatigue properties relative to their strengths. Ti-10-2-3 is weldable, but electron-beam welding is recommended as plasma and TIG welding can result in poor ductility and toughness (Messler, 1981). Alloys such as Ti-15-3 and β-21S are readily weldable.

Alloy Ti-15-3 was developed to improve strip producibility, cold formability, and the ability to heat treat to high strengths. It has excellent cold-forming characteristics for simple forming operations such as brake forming or forming into shapes. However, for more-complex forming operations, such as tube bending, stretch and bulge forming, where triaxial stresses are developed, forming difficulties can be encountered.

Alloy Ti-10-2-3 is the most highly used of the β alloys, has excellent fatigue properties, and has moderate fatigue crack growth rate characteristics. The most significant application of Ti-10-2-3 is on the landing gear of the Boeing 777, which results in a significant weight savings compared with 4340 steel and eliminates the potential for stress corrosion cracking associated with steel.

Timetal 21S has good high-temperature properties, with creep properties superior to that of Ti-6-4 (Fanning, 1993). Applications on the Boeing 777 are in the engine nacelle and in areas where exposure to hydraulic fluids at elevated temperatures can occur (alloy 21S is uniquely resistant among titanium alloys to hydraulic fluids used in commercial aircraft).

Low-Cost Alloys

Two low-cost titanium alloys have recently been developed. Timetal 62S (Ti-6Al-1.7Fe-.1Si) was developed as a low-cost replacement for Ti-6-4 for the automotive industry (Bania et al., 1993). Since iron is a much lower-cost alloying addition than vanadium, the use of an expensive master alloy was eliminated. The other low-cost alloy, Timetal LCB (Ti-6.8Mo-4.5Fe-1.5Al), developed for automotive springs, takes advantage of low-cost alloying additions by using a low-cost ferro-moly alloying addition (Bania, 1994). This alloy can be heat treated to strengths in excess of 200 ksi (1,378 MPa) with reasonable ductility. The properties of both of these alloys indicate that they may be appropriate for airframe applications. Timetal LCB is presently being studied as a high-strength fastener alloy.

METAL-MATRIX COMPOSITES

Much of the early work on metal-matrix composites (MMCs) involved aluminum-matrix alloys. Recently, matrices based on titanium alloys and intermetallics of titanium and aluminum (γTiAl) have received much interest. There are a variety of types and morphologies of reinforcements used in MMCs, principally high-melting-point ceramics, such as SiC or Al_2O_3, in the form of discrete whiskers, particles, or continuous fibers.

The major benefit of MMCs over monolithic alloys is their higher strength, elastic modulus, and fatigue crack initiation resistance at the expense of lower toughness. The major emphasis in research has been to achieve improved ductility and toughness in discontinuously reinforced MMCs and improved toughness in continuously reinforced MMCs with no loss in strength. Unfortunately, the costs of producing MMCs are high. In MMCs with continuous reinforcement, key issues include cost, processing, and producibility of useful shapes.

Continuously reinforced MMCs provide the greatest strength and stiffness at premium cost. Landing gear on advanced aircraft can use continuously reinforced MMCs for reduced weight and increased environmental resistance. Other candidate applications include supersonic aircraft skins and engine structures where high-temperature strength is required.

Discontinuously reinforced MMCs, containing whiskers or particles, provide increased strength and stiffness, but at higher costs than unreinforced metals. They can find applications in lightly loaded, stiffness-critical airframe components where enhanced fatigue or fracture resistance is not a necessity. Examples include inertial guidance systems, rudders, escape hatches, and aircraft hydraulic systems.

MMCs with continuous reinforcement have a problem with fiber-matrix compatibility, fiber cost, fiber size, and fiber-coating technology. There are also unresolved issues associated with consolidation technology, the cost of production, and manufacturing, including post-fabrication shaping, forming, and machining, as well as the establishment of design properties.

Whisker and particulate MMCs need specially designed dies for primary processing. Achieving a uniform dispersion of particles and producing a controlled or reduced whisker or particulate size is difficult and processing costs are high.

The major barrier to the use of MMCs has been their high cost. Other barriers include the lack of standardization of mechanical property measurements and difficulty in machining. Process development and standardization are needed for both continuous and discontinuous MMCs. Other constraints include low fracture toughness and poor, short transverse mechanical properties. Because of these constraints, the committee foresees niche applications but not major use of MMCs in next-generation commercial transport airframes. The most likely first application of MMCs in commercial aircraft is in engine applications; however engine applications are not within the scope of this study.

TRENDS IN PROCESSING

Forming

A range of metallurgical forming processes are employed in the production of commercial aircraft. These include both cold-forming and hot-forming processes. The process used depends on the characteristics of the alloys and on the amount of deformation required. Two forming processes of particular importance for next-generation aircraft will be age forming and superplastic forming and are described later in this chapter.

Age forming utilizes the metallurgical stress relaxation phenomena that occurs during the artificial aging or heat treatment of aluminum alloys. The age-forming process can be performed on any of the heat-treatable aluminum alloys in the 2XXX, 6XXX, and 7XXX series. Age forming offers a potential solution to many of the problems encountered when conventional cold-forming processes are applied to integrally stiffened, complex shaped parts. Stress relaxation occurs during the age-forming process to convert elastic strain into retained deformation for simple and compound contour shapes. Production use of age forming has occurred primarily on wing skins and stringers, with experience on B-1B upper and lower skin panels, Gulfstream IV compound curvature upper wing panels, Airbus A330 and A340 upper wing panels, and iso- and ortho-grid patterns for Titan IV booster skirts.

Uniform pressures are applied at the required aging temperatures using bagging and autoclave techniques. Both peripheral and total bagging methods have been employed successfully in the development and qualification of the age-forming technique. Process cycle temperatures and times can be adjusted and typically correspond to MIL-HDBK-5F aging practices. Post-forming property evaluations in 2419, 7150, and 8090 Al-Li alloys have demonstrated no alteration in mechanical properties due to the age-forming process. Parts with complex thickness changes, cut-outs, pad-ups, and stiffeners have been formed successfully using age-forming procedures.

Tooling development involving primarily the issue of "over-form" or spring-back is the major factor to be understood in forming new parts. For integrally stiffened concepts, 46 cm x 61 cm (18 in. x 24 in.) test panels are normally used to aid in both design and tooling concepts. Concave and convex contoured panels with grid patterns have been fabricated using age forming in a variety of high-strength 7XXX- and 2XXX-series aluminum alloys. Radii of panels fabricated to date have been on the order of 100–150 cm (40–60 in.) in diameter, although 50–76 cm (20–30 in.) in diameter appears to be feasible with proper development.

Near-Net-Shape Forging and Casting Processes

Aircraft alloys are producible by conventional methods and much has been written regarding techniques and capabilities. The continuing issue is that of cost to produce usable part configurations. For example, the use of titanium has been limited by raw product costs (from 3–10 times that of aluminum or steel) and processing difficulties. Since the early 1970s, government and industry research has focused on low-cost processing methods for both military and commercial aircraft, with much of the focus on net-shape processing methods that eliminate (or reduce) costly post-processing operations such as machining (AFML, 1973; NRC, 1986). Processes that take advantage of this approach include net-die, hot-die, or isothermal forging, and net-shape premium-quality castings.

Forging Processes

Conventional forging technology produces complex die forgings, by press or hammer techniques, to configurations with average "buy-to-fly" (B/F) ratios[1] of about 7:1. Smaller, simpler forgings (<323 square cm or 50 square in. plan view area) can be produced with B/F ratios of 2:1 or less (precision forgings), while larger, complex forgings have greater B/F ratios, sometimes as high as 25–30:1 or greater. Subsequent machining processes to reach the desired final configuration can be very expensive, especially in the case of titanium, where forging stock can cost as much as $25 per pound.

[1]The buy-to-fly ratio is a measure of materials utilization. It is defined as the ratio of the average weight of material purchased to the final component weight.

Numerous U.S. Air Force and industrial technology development programs have been conducted to establish a technology base for hot-die (isothermal) forging of titanium alloys closer to net size (Leodolter, 1982). These efforts resulted in the development of a technology base for producing complex near-to-net titanium with B/F ratios of 3:1 or less.

Although widely used for jet engine components, because of the larger size and asymmetrical shapes required for airframe parts, isothermal forging has not been extensively used. Furthermore, it does not appear that the technology will find much use in next-generation aircraft for several reasons:

- Building a complex, high-temperature die system is very costly and time-consuming, and delivery schedules of first parts generally will not allow time to implement the process. It is likely that production runs will not be sufficiently long enough to amortize the cost of the dies. The time to set-up for a production run would not be justified, based on the limited number of articles that are required.
- Competing processes, particularly castings, will allow faster production of the same parts, closer to final configuration, and in many cases to a more structurally efficient configuration than is producible by any combination of forging and machining.
- The major driver for net-shape processing has been the high cost of machining, especially for titanium because the material costs are significantly higher and because titanium is more difficult to machine than aluminum or steel. However, with the development of advanced machining tools, numerical controls, and innovative cutter technologies, machining of titanium is no longer the high-cost operation it has been in the past.

Although forgings will continue to be a mainstay structural product form for the next-generation aircraft, in light of the above it is likely that they will be either conventional forgings or forged block.

Casting Processes

Historically, strength and toughness levels and consistency in quality of cast parts have been inferior to wrought fabricated products. Consequently, castings have not been used in many airframe primary structure applications. However, both aluminum and titanium castings are finding significant new application in aircraft, especially military systems (AGARD, 1991). The increased application of castings has been fueled by advances in process technology and the need to reduce manufacturing costs.

Aluminum castings have been produced and utilized in a wide range of airframe components, including pylons, bulkheads, vertical stabilizers, and canopy frames. The most common aluminum casting alloys include A356, A357, and A201. Aluminum casting alloys generally exhibit lower allowable design stresses than wrought alloys which is attributable to the wider variability in properties compared with wrought products because of the persistent occurrence of porosity. Improved aluminum casting alloys, advances in understanding microstructural control, and improvements in sand- and investment-casting processes have resulted in significant increases in the use of aluminum castings, particularly by Airbus.

Cast titanium components have been used successfully in both engines and airframes for many years. There are numerous applications of titanium castings for structural applications such as frames and nozzles. The predominant titanium alloy for cast parts in aerospace structures is Ti-6-4, but several other alloys, including Ti-6-2-4-2s, Timetal-1100, and Ti-15-3, have also been investigated.

In airframe applications, titanium castings (particularly Ti-6-4) are being used in secondary or nonstructural applications with casting factors[2] of 1.0–2.0, depending on the criticality of the application. As casting technology has matured, and as hot isostatic processing (HIP) has become an integral part of the casting process for titanium, larger, more complex cast shapes with significantly improved structural properties have been realized.

The advances in casting processes, particularly HIP, the improving performance database, and prior history in engine applications led the Boeing/Lockheed team to commit to produce the F-22 side-of-body rib assembly (figure 3-1), a primary structure/fracture critical part, from two Ti-6-4 castings with a casting factor of 1.0. Benefits in weight and cost were realized because of design refinement, joint elimination, and minimization of secondary processing (such as machining). This initial application of castings on the F-22 has led to many additional applications which total more than 50 parts, many in fracture-critical applications. The success on the F-22 should provide impetus to the commercial airframe industry to pursue similar applications.

The issues that need to be addressed before wider use can be made of critical structural castings include:

- Elimination of casting factors. Extensive tests are required to determine static properties and durability and damage tolerance characteristics that permit reduction or elimination of casting factors (currently 1.3–1.5). The implementation of process controls and the development of property databases will allow the minimization or elimination of casting factors.

[2]Casting factor is an additional margin of safety multiplier imposed on cast components because of the historical inconsistencies in casting processes and component quality.

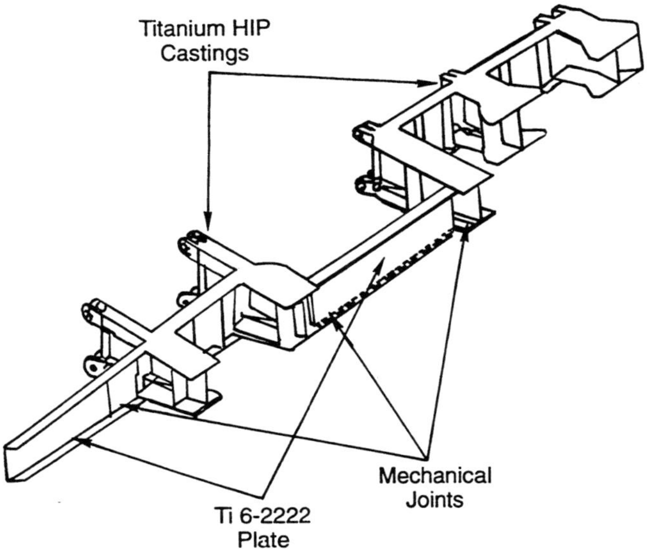

FIGURE 3-1 HIP castings applications—F-22 wing-to-body rib castings. Source: Clark et al. (1995).

- Optimization of casting parameters. The introduction and implementation of computer simulations of the casting processes (including solidification and heat transfer models) will contribute to improved designs and fewer defects. By using process simulations, changes in casting configurations and mold designs can be more efficiently achieved to accelerate final part production.
- Rapid prototype development. Rapid prototyping technologies encompass a variety of methods to directly generate three-dimensional models from computer-aided design (CAD) workstations. The solid models can be used as patterns for investment castings and to verify dimensional conformance on final parts.

Further advancements in the casting industry will extend the application potential and cost benefit significantly. One such advancement is rapid prototyping methods, such as using stereolithography to produce first articles in a short period of time. Stereolithographic technology may also lead to low-cost mold fabrication techniques. Finally, significant progress is being made by industry, as well as academic, researchers in modeling of the casting process (NRC, 1995a). As the technology matures and extends into more complex shapes, further benefits will be realized, including reduced time to first part development, increased shape complexity, and improved casting integrity.

In summary, castings will experience increasing use for primary structure on airframes because of the improvements in the capability of the casting industry to produce complex, high-integrity castings. Future advancements inclusive of additional alloys, rapid prototyping, stereolithography, process modeling, and welding will further extend their use. Increased confidence will finally reduce or completely eliminate casting factors.

High-Speed Machining of Low-Residual-Stress Parts

Semifinished material is not a major contributing factor to the acquisition cost of advanced aircraft. Accordingly, it is possible to trade off some efficiency in material utilization in order to achieve significant decreases in assembly costs. In many cases, recent advances in high-speed machining have out-paced developments in net-shape processing (Smith, 1994a).

High-speed machining is an attractive alternative to conventional machining. Cutting heads move at significantly higher speeds, and since cutting forces are lower, smaller cutting radii and thinner sections can be realized. Gauges obtainable are thin enough that complex, built-up sheet-metal assemblies can be considered as replacements for conventional machined components without weight penalties. Even though material utilization factors can be reduced, subsequent assembly costs will be reduced as well. Overall surface quality is better than that obtained with conventional machining. Existing machining equipment can be used for some high-speed machining operations, but to take full advantage of this technology, some capital investment will be required.

In combination with computer-controlled, multiaxis machining heads and emerging CAD software, many complex part configurations can now be more fully considered. Biaxially stiffened concepts with integral web channels produced by high-speed machining show encouraging structural efficiencies at affordable manufacturing costs.

Superplastic Forming and Diffusion Bonding

Superplastic forming (SPF) is a hot forming process wherein, due to the metallurgical structure of the alloy (very fine grain size) and the processing temperature, the material exhibits superplasticity (as much as 1,000 percent elongation) within a range of low strain rates. In some cases, SPF is combined with diffusion bonding (DB) wherein parts are held in intimate contact during the high-temperature forming process, allowing a high-integrity bond to form because of interdiffusion across the interface.

Aircraft components that can be produced using superplastic forming include complex, secondary structural parts that require high local deformations in one forming operation. This section describes SPF and DB of aluminum and titanium alloys.

Aluminum Alloys

Aluminum alloys do not exhibit superplastic properties under conventional processing conditions. Special thermomechanical treatments are usually necessary to obtain the desired grain structure necessary for superplastic deformation. Aluminum alloys (e.g., 2004, Supral 100, and Supral 220) were developed especially for their superplastic properties.

Aluminum alloy 2004 (Supral 100) is a medium-strength alloy with mechanical properties similar to 6061 and 2219 and is normally used in lightly loaded or nonstructural applications. Supral 100 components have been produced for over 80 different aircraft, including the Airbus A340, Aerospatiale's ATR, and the Boeing 777, and are in service in many countries around the world. For example, the wingtip light housing for the Boeing 777 is fabricated using SPF processing. The European Fighter Aircraft Program is considering many SPF parts, including sine-wave spars, auxiliary power unit (APU) shear wall, tank shear walls, doors, boxes, fire walls, and outlets. Based on these service experiences, SPF of aluminum components should see increased use in commercial aircraft.

Fabrication methods that combine SPF and DB, as used for titanium alloys, are not commercially available for aluminum alloys. The effective use of DB is hindered by the tenacious oxide film present on all aluminum alloys. Development programs directed at overcoming this problem are underway but have not yet been successful.

Titanium Alloys

SPF and DB technologies as applied to titanium alloys are essentially mature. Both are currently being used on the F-15E aircraft with significant benefit. Strength properties of the SPF/DB component are equivalent to that of wrought alloys. However, the surfaces being joined must be very clean or strength will be reduced.

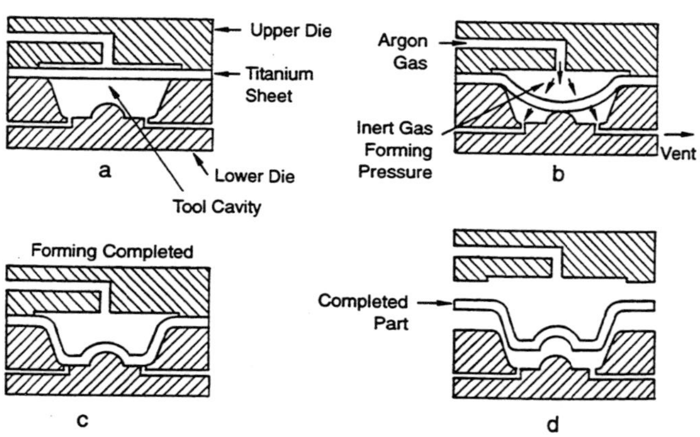

FIGURE 3-2 SPF process. Source: Gerber (1990).

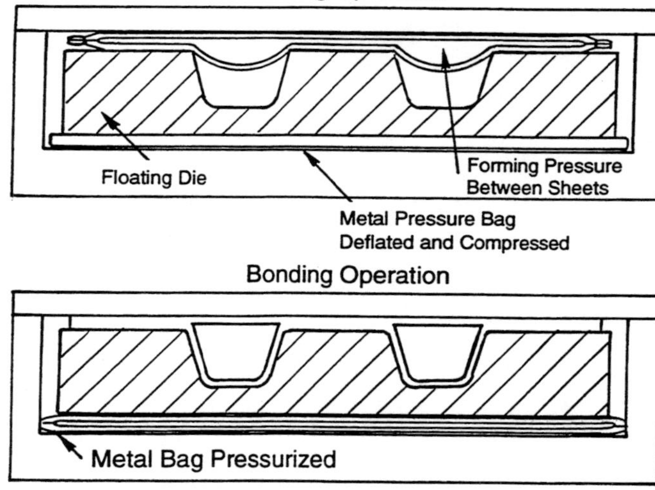

FIGURE 3-3 Two-sheet SPF/DB process. Source: Gerber (1990).

Numerous techniques have been developed that utilize the SPF and DB process. Several are shown in figures 3-2, 3-3, and 3-4 which depict the basic SPF process, two-sheet SPF/DB, and four-sheet SPF/DB, respectively. One can see from the figures that the process is versatile and can be adapted to forming a myriad of complex cross-sections in flat, curved, or compound contours. Currently, there are at least 10 companies in the United States with the capabilities to produce SPF and SPF/DB titanium parts or structures. Parts as large as 122 x 488 x 46 cm (48 x 192 x 18 in.) can be produced by this extensive industry base.

The large research and development effort, extensive industrial capacity, and current applications notwithstanding, the basic process has a number of technical issues that must be addressed before it will gain general user acceptance. These issues include:

- long cycle times;
- requirements for part forming prior to bonding operation to meet contour tolerances;
- the effect of cooling rates on mechanical properties;
- quality of bonding, especially in edgeband areas;
- inspection methods and acceptance and rejection criteria;
- cost, durability, and surface-finish retention of tooling materials; and
- reduced fatigue life (at least 10 percent) for current two-sheet and four-sheet configuration approaches.

Many of the above issues are being addressed in ongoing developmental activities throughout the aerospace industry. SPF and SPF/DB have the potential to reduce the cost of a number of parts and assemblies. SPF will continue to be a

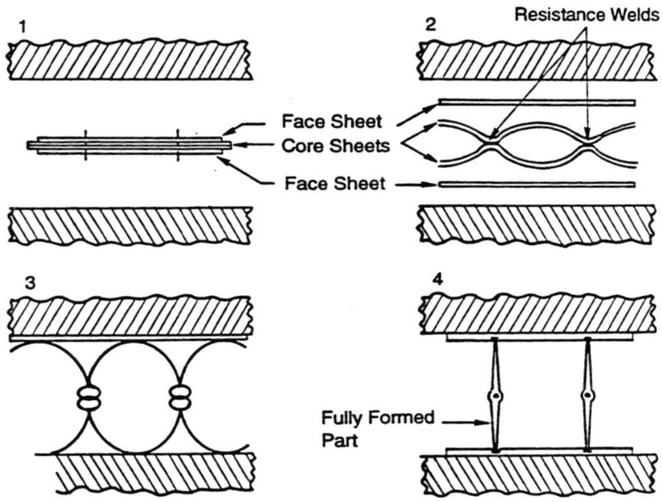

FIGURE 3-4 Four-sheet SPF/DB process. Source: Gerber (1990).

candidate process for application whenever design requires complex shapes with compound curvature.

SUMMARY

Aircraft alloy materials, including high-performance aluminum, high-strength steels, and titanium alloys, have been advancing steadily with each new aircraft model. Significant progress is being made, primarily through incremental improvements to existing alloys, in developing alloys with improved strength, toughness, corrosion resistance, and producibility.

The desire of aircraft designers to improve durability and save weight has led to the development of new aluminum alloys that provide improved combinations of specific strength, durability, and damage tolerance. Most often, the newer alloys are variants of older alloys, but with tighter controls on chemistry and processing parameters.

Al-Li alloys have seen limited commercial use. The Al-Li alloys as a group have attractive fatigue properties; are amenable to superplastic forming; display moderate to good weldability; and can be chemically milled, bonded, anodized, clad, and painted. However, they often display considerable anisotropy, are typically more costly than conventional aluminum alloys, are more susceptible to surface oxidation, and are prone to warping during quenching. Al-Li alloys are more difficult to process than conventional aluminum alloys, and their properties can be greatly influenced by relatively minor processing variations.

Improved Ni-Co, low-carbon steels (most notably Aermet 100 and AF1410) have been developed with outstanding combinations of high strength, high toughness, and corrosion resistance. These alloys have found use in landing gear. The strength of these new steels cannot increase to significantly higher levels without corresponding decreases in ductility and toughness. However, increased strength can be achieved while keeping toughness at levels that may be acceptable for landing gear applications for civil aircraft.

The primary reasons for using titanium for aircraft applications include weight savings, high operating temperature, space limitation, corrosion resistance, and composite compatibility. The committee believes that titanium alloys will continue to be developed with improved performance in terms of temperature resistance and strength/toughness combinations. Lower-cost titanium alloys have the potential to further increase the utilization of titanium alloys in production applications.

Despite the potential for significant performance benefits, the committee does not foresee significant use of metal-matrix composites in the airframe of next-generation transports because of their high cost, lack of standardization of mechanical property measurements, and difficulty in processing and machining. Process development and standardization are needed for both continuous and discontinuous metal-matrix composites.

Manufacturing process development has been emphasizing low-cost approaches such as net-shape processing (casting and forging), improved forming methods (including age forming and superplastic forming), high-speed machining, and superplastic forming and diffusion bonding.

4

Polymeric Composite Materials and Processes

The application of polymeric composites has been an evolutionary process, with increased use as materials and processing technology matured and program needs dictated their use. Glass-reinforced composites, in the form of thin-facesheet honeycomb sandwich constructions, have been in general use for secondary structures (i.e., wing-to-body fairings, fixed-wing and empennage cover panels, and secondary control surfaces) on commercial transport aircraft since the mid-1960s.

During the 1970s, the commercial availability of carbon and aramid fibers and the uncertainty in fuel supply and costs provided an impetus for the development and application of structural composites for airframe applications. NASA conducted technology development and flight-service programs to encourage the use of composites in commercial production applications. Components, summarized in table 4-1, were designed, certified, and used in airline service. These flight-service programs were critical in developing design methods and in building confidence in composite applications. A summary of flight-service and long-term exposure tests has been recently published (Dexter and Baker, 1994).

Carbon/epoxy, aramid/epoxy, and aramid-carbon/epoxy and glass-carbon/epoxy hybrid composites were first used on a production scale in the early 1980s for the generation of aircraft that included Boeing 757, 767, and 737-300; Airbus A310 and A320; and McDonnell Douglas MD-80 series. Applications included secondary structures such as fairings, fixed-wing and empennage cover panels, and engine cowlings, as well as primary flight controls such as ailerons, elevators, rudders, and spoilers. The materials used for these components included largely unmodified amine-cured epoxy resins (e.g., TGMDA/DDS) reinforced with aramid (Kevlar® 49), carbon (e.g., Toray/Amoco T-300, Hercules AS-4), and E-glass fibers. Constructions were generally facesheets co-cured or secondarily bonded to composite honeycomb core.

The first production application of composites on primary structure was in the late 1980s on Airbus A320 empennage components. The construction used was an integrally stiffened carbon/epoxy laminate skin fabricated from materials similar to the first-generation materials previously used for secondary structure and primary flight controls. The further development of carbon fibers with improved strength and modulus (e.g., Hercules IM7 and Toray T-800H) and high-performance and toughened matrix polymers has led to application on the Boeing 777 empennage to expand the primary structural applications.

The factors currently driving new materials applications on commercial aircraft will place added emphasis on design simplification, low-cost processing, and durability and maintainability. The committee believes that continuing developments in toughened thermosetting polymers and high-performance thermoplastics will contribute to these goals. Perhaps more important is the need to validate

TABLE 4-1 NASA Flight-Service Components

Aircraft Model	Component	Material	Number of Parts	Entered Service
L-1011	Underwing fairing	Aramid/epoxy	18	1973
L-1011	Aileron	Carbon/epoxy	8	1982
DC-10	Upper aft rudder	Carbon/epoxy	15	1976
DC-10	Aft pylon skin	Boron/aluminum	3	1975
DC-10	Vertical stabilizer	Carbon/epoxy	1	1987
727	Elevator	Carbon/epoxy	10	1980
737	Spoiler	Carbon/epoxy	108	1973
737	Horizontal stabilizer	Carbon/epoxy	10	1984

innovative low-cost manufacturing processes. This chapter provides an overview of key composite material and process technologies for next-generation transports.

COMPOSITE MATERIALS DEVELOPMENT

Early efforts to develop composite primary structures for military and commercial aircraft were limited by the need to provide damage resistance (i.e. the damage created during an event such as impact) and damage tolerance (i.e. the effect of a given state of damage on structural performance). First-generation composites, with matrix systems that were generally relatively brittle thermosetting epoxies, had low through-thickness strength and were susceptible to through-thickness damage, particularly delamination, from transverse loading, such as that which occurs in impact events. Structural schemes were developed to improve the structural damage tolerance manifested as improved residual strength after low velocity impact. This included the use of lower stiffness laminates as the skin with high-stiffness doubler planks and stiffeners. However, the manufacturing cost and weight of the resulting structure made such composite primary structural applications difficult to justify for commercial aircraft. This led to a need to develop "toughened" composite systems.

Improved "toughness" in a composite system was originally defined by compressive residual strength after an impact event. Work in the area showed that the resistance of the composite to delamination is a key property related to this structural capability. A large amount of work to develop a characterization capability has resulted in the definition of delamination fracture toughness and tests to determine this property (ASTM, 1994).

Over the past 15 years, the aircraft and materials industries have investigated matrix polymers, with the goal of improving the damage resistance and damage tolerance of structures made from these systems. Toughened thermoset and thermoplastic polymers, described in the following sections, with improved values of delamination fracture toughness and resulting improvements in the structural parameter of compression strength after impact resulted from this work. The committee believes that future work will focus on the development of systems compatible with innovative processes described later in this chapter and the optimization of composite properties to provide resistance not only to delamination, but also to other forms of damage such as fiber damage and matrix cracking.

Toughened Thermosets

Boeing catalyzed the development of toughened composite systems for commercial aircraft in 1982 with the release of a preliminary specification (XBMS 8-276) outlining performance goals for a primary structural composite material. Since that time, advances in matrix polymer technology have enabled the development of composite systems with improved delamination resistance.

One approach to matrix toughening is through polymer "alloying" of brittle resins with tougher thermoplastic or rubber systems. This can take the form of polymeric blends with discrete second phases, interpenetrating networks, or random or block copolymers.

Also, composites can be made more resistant to certain types of impact damage through architectural tailoring. An example is the use of tough, unreinforced polymers between the brittle, high-strength composite plies to increase delamination resistance. This approach was used in some of the more successful candidate systems (Krieger, 1984; Hirschbuehler, 1985; Masters et al., 1986).

A combination of the blend and interleaf approaches was the key to finally meeting property goals for both impact and temperature resistance (Chu et al., 1987; Odagiri et al., 1991). These systems use tough modifier particles of controlled size to create a tough resin-rich interleaf layer at the interply region. The base polymer is a thermoset thermoplastic blend. Toray T-800H/3900-2, the system qualified for the Boeing 777 empennage skins, stringers, and spars as well as floor beams, uses this approach.

Thermoplastics

In recent years there has been an increased interest in continuous-fiber-reinforced thermoplastic-matrix composites (Cogswell, 1991, 1992). High-performance thermoplastic polymers provide combinations of toughness and temperature resistance that cannot be attained by most thermosetting matrix systems (NRC, 1987). While the toughness of thermoplastic matrix polymers may improve composite delamination resistance, the outstanding properties of high-performance thermoplastic polymers have not translated well into improved composite performance. For example, depending on the laminate configuration and loading, a thermoset composite with a hole could sustain a higher ultimate load than a similar thermoplastic composite because of the effect of local damage redistributing stress concentrations around the hole. This can be particularly important in loaded holes (e.g., fastened joints).

Thermoplastic matrices can be processed by simply heating above the softening point and applying molding pressure. Thermoplastics can be repeatedly heated and reconsolidated without degradation of properties. The properties and processing characteristics of high-performance thermoplastics are a result of their high molecular weight (>20,000) and aromatic molecular structure.

Thermoplastic-matrix composites offer potential manufacturing advantages over thermosetting systems, including fast cycle times, virtually unlimited shelf-life, and the ability to reform or reconsolidate. However, thermoplastic systems are rigid at ambient shop temperature and have no tack and limited drapeability to allow conformance to tool contours, making ply orientation difficult. Also the high melt temperature and high melt viscosity require thermoplastic systems to be processed at temperatures (>300°C or 570°F) and pressures significantly higher than those required for thermosets—limiting the configuration and size of parts to press-size capabilities. Large parts have been produced in autoclaves, but the cycle-time advantages of thermoplastics are lost due to heat transfer limitations in autoclave processes.

Innovative processes are being developed to take full advantage of thermoplastic systems. These include diaphragm forming, pultrusion, and in situ consolidation. These processes are described later in this chapter.

Next-Generation Systems

Past investigations of toughened composites have focused on evaluation of impact damage visibility and residual strength. Recent work (Dost et al., 1993) indicates that a more complete understanding of the performance of toughened systems would result from investigation of both damage resistance (i.e., damage created during some event such as impact) and damage tolerance (i.e., the effect of a given damage state on structural performance). There have been results that indicate that the residual strength of composite systems with toughened thermoset matrices are more sensitive to large notches than brittle matrix systems (Walker et al., 1992). The increased notch sensitivity for toughened matrix systems is attributable to increased fiber damage and matrix cracking that occurs in systems with high delamination resistance. The committee believes that, in the future, materials optimization efforts will take a balanced approach that considers both small-scale (including most impact threats) and large-scale (including penetrations and severed elements) damage (Dost et al., 1993).

Although toughened systems have been successfully applied in production applications, these systems are relatively expensive and were not developed to be compatible with innovative process technologies such as tow placement, resin transfer molding, or resin film infusion. In the future, developments will focus the effect of processing on toughening mechanisms and on low-viscosity, single-phase or two-phase systems where constituents are soluble at processing temperatures.

TRENDS IN PROCESSING

Although high costs for raw materials have been blamed for the slow growth of composites in the marketplace, material costs actually account for only 8–10 percent of the overall cost of composite components (DeVault, 1993). In fact, manufacturing costs are the single largest contributor to overall costs (JTEC, 1994). While the development of composites for aerospace applications has traditionally been driven by performance, cost has assumed increasing importance during the past several years. Thus, a primary criterion in the development of manufacturing processes for the next generation of commercial transports has been the potential for low-cost production of components. The committee believes that the trend to develop low-cost production processes will be continued for the foreseeable future. Trends in low-cost composite processes are discussed in this section.

Resin Transfer Molding

Resin transfer molding (RTM) is a closed-mold process that allows the fabrication of component geometries ranging from simple, low-load carrying structures to complex, high-performance hardware (Steenkamer et al., 1993a,b). This process is illustrated in figure 4-1. Preforms fabricated from dry fibrous glass or carbon reinforcements in the form of woven cloth, stitched broad goods, or randomly oriented mats are placed in a matched tool and closed via a press or clamping fixture. Vacuum can be applied to the tool to aid in the removal of trapped air and moisture from the preform, to lower the required resin injection pressures, and to ensure proper tool seal prior to resin injection. A thermoset resin, either a one-part or two-part catalyzed system, is heated to decrease viscosity and injected through an injection gate into the cavity at pressures typically ranging from 30–60 psi. The resin flows inside the tool cavity, impregnates the reinforcement, and is cured at elevated temperatures. The keys to successful RTM are high-quality tooling, preforms with consistent fiber volumes, and strategic placement of the injection gate and vents to optimize resin flow and remove trapped air.

RTM is a relatively flexible process, with its principal advantage being the ability to produce large, complex geometries with high-fiber volumes and structurally efficient designs. Both the advantages and limitations of RTM are summarized in table 4-2.

RTM employs a variety of fiber forms, including randomly oriented mats or structurally efficient textile preforms containing glass, carbon, and aramid fibers. Processes used to produce fiber preforms are discussed later in this chapter.

The thermoset resin systems most commonly used in the RTM industry include polyesters, vinyl esters, and epoxies. There has also been limited RTM work performed with bismaleimide systems. Epoxies and bismaleimides are generally used for aerospace applications because of their mechanical properties and environmental resistance. Low resin viscosity (~200–400 cp in the tool) and long pot life, depending on part size and injection method, are key attributes to

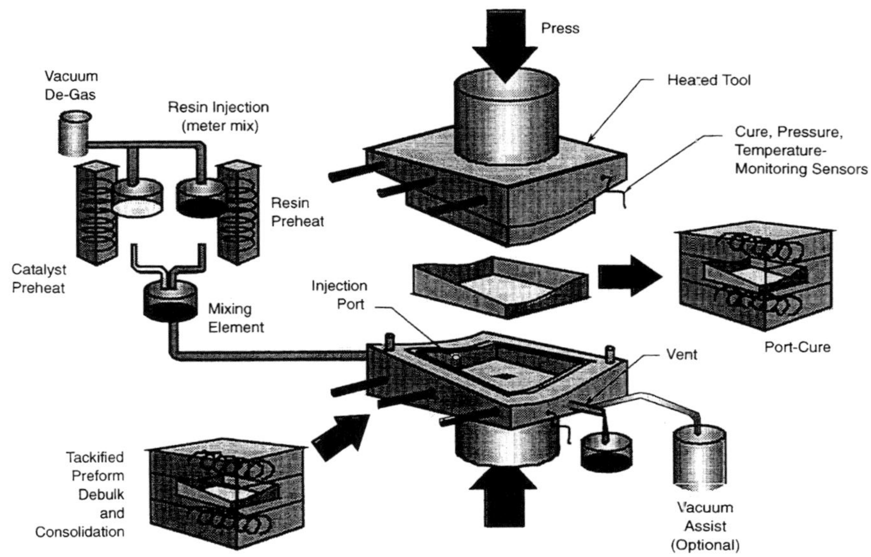

FIGURE 4-1 RTM process. Source: Blanton et al. (1995).

consider when selecting an RTM system. Other factors to consider when selecting a resin system include cost; chemical resistance; moisture absorption; shrinkage; and compatibility with coatings, adhesives, or fiber sizing.

Future RTM structures will compete in cost while attaining the performance necessary for application to commercial aircraft. Factors that will reduce the cost of the RTM process (Fowler and Phifer, 1993) include:

- low-cost resins and hardeners;
- optimized preform design and fabrication;
- automated multipurpose molds;
- proper selection of fast-curing resins for reduced cure time;
- mold design for easy mold clean-up; and
- molded-in cut-outs, holes, and trim to eliminate finish machining operations.

These cost-reduction measures, coupled with emerging high-performance resin systems, will allow the RTM process to be utilized for the fabrication of primary structural composite parts.

A process that is related to RTM is vacuum-assisted resin infusion in which resin is pulled by vacuum through a high-permeability membrane placed on top of a preform. The process features low-cost, one-sided tooling and vacuum-bag technology. This process was developed by Seeman Composites for large composite structures for marine (e.g., ship hull sections) and infrastructure applications (e.g., bridge decks) consisting of E-glass-fabric preforms injected with vinyl ester resins. This technology has the potential to manufacture near-net-shape components at costs approaching $10 per pound of structure. For aircraft applications, the higher-temperature resin systems as mentioned above are required. Process automation and control is a key requirement to ensure that parts can be fabricated with reproducible quality and performance. New developments in flow sensors will be required to ensure complete wetting of fiber within tows, as well as the detection of macroscopic defects (i.e., dry spots).

TABLE 4-2 Advantages and Limitations of the RTM Process

Advantages
Matched die molding (improves dimensional control, stability, and surface finish)
Reinforcements can be tailored in critical areas (reduces weight)
Large, complex components can be produced as a single part (reduces weight and assembly costs)
Near-net-shape parts can be produced (reduces secondary trimming and assembly costs)
High fiber volume (55–60 percent) parts that include features such as molded-in metal inserts, lightweight core materials, and thickness variations can be produced
Two-part resin systems that do not need refrigeration (e.g., epoxy and vinyl ester) are generally used
Capital equipment and tooling can be relatively inexpensive compared with other closed-mold processes
Time-consuming debulking steps are eliminated
Limitations
Lower production rates than other molding processes
High-performance aircraft resin systems require further advances in process development, materials characterization, and physical property optimization.
Labor required for placing reinforcements in molds can be high
Initial mold costs higher than for one-sided molding processes
Control of voids and process variability

SOURCE: Blanton et al. (1995)

Resin Film Infusion

The resin film infusion (RFI) process is similar to RTM processes discussed in the previous section. In this process, dry-fiber preforms are placed into the mold with precast or extruded resin films as shown in figure 4-2. Controlled heat and pressure are applied, generally in an autoclave. Infusion parameters are established by a combination of experimental and modeling approaches. This process results in high-quality, near-net-shaped parts that require only clean-up and limited trim.

For thick composite structures, such as the composite wing discussed in chapter 2, the RFI process eliminates limitations imposed by out-of-refrigeration restrictions for perishable, preimpregnated laminated materials with thermoset matrix resins.

Dry-fiber preforms used in the RFI process can be made quickly and therefore economically from commonly used filaments. Recent developments in the area of wing structure utilizes stitching to provide fibers that are transverse to the primary reinforcing filaments (Markus and Palmer, 1991). The stitching creates a preform that retains its shape during handling, will remain sufficiently compliant for insertion into the mold cavity, and as discussed in chapter 2, provides through-thickness reinforcement to improve delamination resistance.

The RFI process has been successfully demonstrated on subcomponent-level structures (Markus et al., 1993). A goal of the NASA Advanced Composite Technology program is to demonstrate this technology on full-scale wing structures by the turn of the century. Integral to the future success of RFI technology will be the ability to accurately predict and control the infusion and cure cycle environment. Modeling techniques to predict exact cure profile sequences for the RFI process and verification techniques are currently under development.

Textile Preforms

Fiber preforms represent a major component of the cost of RTM and RFI structural composites. Preforming relates to the various processes available for converting the dry, unimpregnated yarn, roving, and fabrics into the complex three-dimensional precursor of the part. Currently, the most common preforming techniques include cut-and-place (hand lay-up), directed fiber spray-up, stamping, and textile preforming. This section discusses textile preforming processes. As interest in RTM has grown among aircraft and aerospace manufacturers, textile processes including braiding, knitting, stitching, and weaving have become increasingly popular for the development of preforms for RTM. Automation of the preform process is essential to reduce the cost of RTM structure.

Braided Preforms

There are two distinct methods for producing braided preforms for RTM—flat and tubular fabric braiding (Skelton, 1989). While both forms have their place in the textile community, tubular fabrics are much more common in the composites industry. These fabrics are formed by laying down yarns onto a mandrel that moves through the center of the braiding machine's cross-section at a predetermined rate. Yarns can be deposited at angles ranging from 10° to 85° relative to the mandrel direction. Also, it is possible, through a process called circular braiding, to incorporate 0° fibers if additional strength and stiffness are needed in the axial direction. While circular braiding dates back hundreds of years, recent advances have made it possible to orient fibers in three orthogonal directions, including both Cartesian (x, y, z) and cylindrical (r, θ, z) reference frames. A variety of machines are capable of manipulating fibers in all these directions (Bluck, 1969; Florentine, 1982; Weller, 1985; Popper and McConnell, 1987; Brookstein, 1991; Spain and Bailey, 1991).

Knitted Preforms

Knitting is a process through which looped yarns known as stitches are interconnected to form a fabric. Extensions

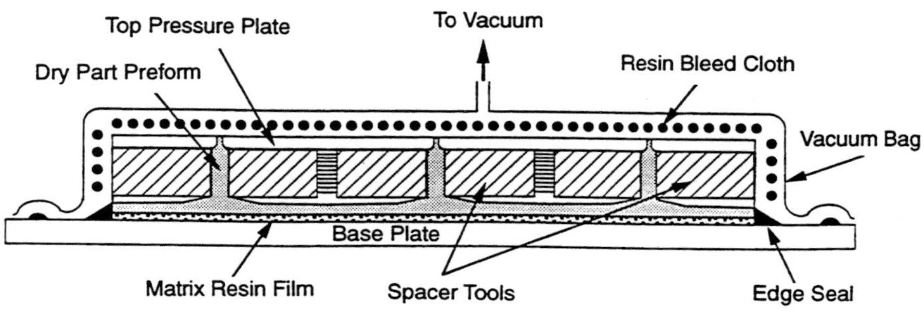

FIGURE 4-2 RFI process. Source: Markus and Palmer (1991).

have been made to the basic warp and weft knitting technologies that allow directional yarns to be incorporated into the fabric structure. In these architectures, the in-plane yarns are the main load-carrying component, while the knitted loop is a secondary structure that serves to hold the in-plane yarns together. The primary types of directionally reinforced knitted fabrics include multiaxial warp knits, weft-inserted warp knits, and stitch-bonded fabrics. These forms are becoming more popular in RTM applications because their construction avoids the crimp imparted to the fibers in braided and woven fabrics.

Stitched Preforms

Unlike other textile processes, which develop preforms by orienting fibers in two or three directions, stitching is primarily used as a means of adding through-thickness reinforcement to a two-dimensional reinforced preform. Stitching can be used locally to lessen the likelihood of delamination as a failure mode and to increase the through-thickness properties of the composite. Also, stitching can be exploited locally, as required, in areas of high interlaminar stresses, and it is possible to vary the pattern, type, or material that is stitched to arrive at the desired performance level (Holmes et al., 1991). The gain in through-thickness response is to some extent at the expense of the in-plane properties and fatigue resistance. Work on the stitching of carbon/epoxy laminates has shown that this loss may be as little as an 8 percent decrease in tensile strength (Ogo, 1987).

Woven Preforms

There are two methods for producing woven preforms. The basic form of weaving involves the interlacing of two sets of perpendicular yarns. Extensions have been made to the basic form of weaving so that it is possible to produce three-dimensional reinforced woven preforms. The two main techniques for developing these preforms are angle interlock and orthogonal weaving. Orthogonal woven preforms contain three sets of mutually perpendicular yarns. Preforms made by this technique have been used for many years to form ceramic-matrix composites (Ko, 1989). Angle interlock weaving is a simple extension of two-dimensional weaving technology where the warp yarns interlace several weft yarns through the thickness of the preform. A variety of preform structures can be formed with the angle-interlock weaving technique (Zawislak and Marden, 1988).

Diaphragm Forming

Diaphragm forming is a technique in which a thermoplastic laminate is held between two deformable sheets know as diaphragms and heated and formed against a tool by hydrostatic pressure (O'Brádaigh, 1990). Only the diaphragms are clamped, so inextensible fibers can be used in the composite laminate. Advantages of diaphragm forming are the degree of complexity achievable and the quality of the consolidated material. Disadvantages include relatively long cycle times (30–60 minutes) and limited deformation of diaphragm materials. Analytical methods are needed to predict the effect of fiber realignment resulting from process deformations on performance.

Diaphragm sheets can be produced from metallic and polymeric materials that are able to survive the high processing temperatures (350–400°C or 660–750°F) associated with thermoplastic composites and substantial deformation without rupture. Superplastic aluminum has been used successfully in forming deep-draw thermoplastic composites. Because of the high processing temperatures, polymeric diaphragms are basically limited to high-temperature polyimide films. Maximum elongation for these films is in the range of 250 percent, which is approximately a factor of 10 less than that of the superplastic aluminum. Efforts continue to develop high-temperature, high-elongation polymeric diaphragm materials which have the potential to reduce fabrication costs by easing part removal.

Diaphragm forming of thermoplastic composite laminates requires pressures up to 200 psi to assure consolidation and complete forming. Therefore, a pressure chamber or autoclave is generally required for forming. The laminate is placed between the two thin diaphragms which are clamped around the edges as shown in figure 4-3.

A schematic of a diaphragm-forming autoclave is shown in figure 4-4. The assembled mold is inserted into the preheated autoclave at the start of the cycle. Upon reaching a

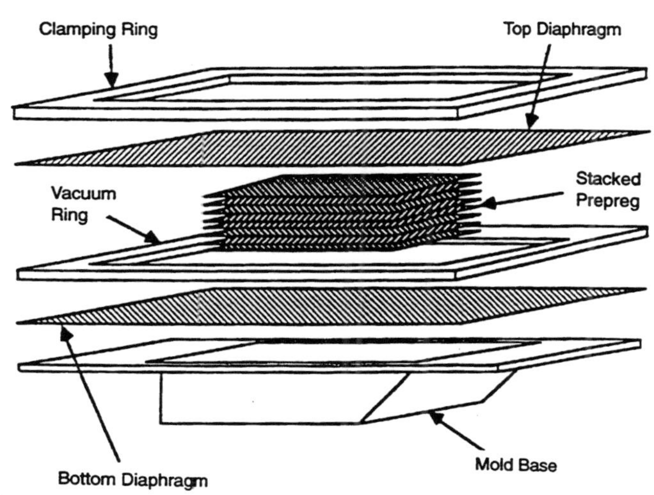

FIGURE 4-3 Mold assembly for double diaphragm forming. Source: O'Brádaigh (1990). With permission from University of Delaware Center for Composite Materials.

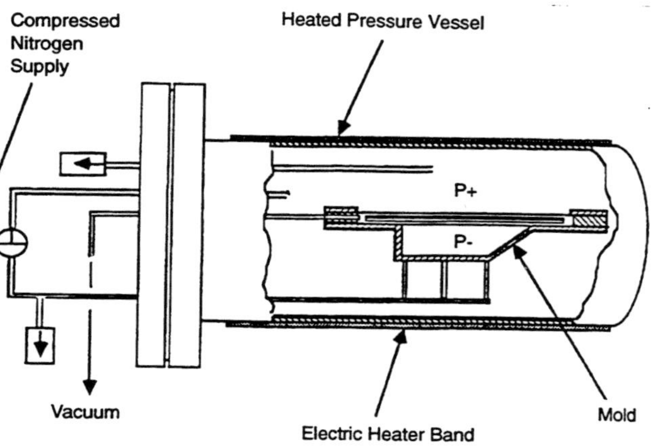

FIGURE 4-4 Schematic of diaphragm-forming autoclave. Source: O'Brádaigh (1990). With permission from University of Delaware Center for Composite Materials.

predetermined, uniform temperature distribution in the laminate, forming is accomplished by applying a greater pressure in the vessel than in the mold cavity. Nitrogen gas is typically used to cool the entire vessel. Differential pressure conditions are maintained during the cooling cycle to ensure the integrity of the formed shape.

Pultrusion

Pultrusion is a relatively inexpensive composites manufacturing technique used to produce continuous structural profiles of constant cross-section, including, for example, beams, channels, angles, rods, and tubes (Fanucci et al., 1991). The process involves pulling matrix-impregnated reinforcement through a heated die to form and consolidate composites of desired cross-sectional shape. Although both thermosets and thermoplastics can be processed using pultrusion, the market is currently dominated by thermosets, as the technology for thermoplastics is much less mature (Larson and Åström, 1991). There is very little material waste with this method (Fanucci et al., 1989). Manufacturing costs can be as low as 15–20 percent of finished product cost, thereby providing high material value.

The advantages of pultrusion include low tooling costs, low labor costs, and the possibility of producing parts with no inherent limit on length. Because pultruded components are pulled through a machined die (usually fabricated of steel), all sides have a smooth finish. Pultrusion also lends itself to tailorability within the limits of the process. The reinforcement can be arranged in a variety of ways—for example, a layer of surface veil, a layer of continuous strand mat, then a layer of unidirectional roving (Winegardner, 1993). Shape complexity is increasing as new forming techniques and new materials are developed. Pultrusion can yield components with multiple cavities, inserts and encapsulations, variable wall thicknesses, and hollow sections (using a cantilevered steel mandrel) (Richard et al., 1983; Goldsworthy and Martin, 1985; Martin and Sumerak, 1987).

The pultrusion process, shown schematically in figure 4-5, begins when reinforcing fibers are pulled from a series of creels through a bath where they are impregnated with resin. The resin-impregnated fibers are preformed to the shape of the desired profile and then enter a heated steel die machined to the shape of the part to be manufactured. The profile is continuously pulled through the die and exits as a constant cross-sectional profile. When the product emerges from the puller mechanism, it is cut to the desired length by an automatic saw.

Several reinforcement forms are available, including continuous roving, continuous strand mat, nonwoven biaxial fabrics consisting of uniaxial plies that are stitched or knitted together, and recently, multiaxial fabrics available with 0°, 90°, and ±45° orientations. While neither roving nor continuous strand mat, on their own, provide adequate performance for most aircraft applications, combinations of all of these fiber types can be used together in pultrusions to provide optimum cost and performance.

The most common resins used in commercial pultrusions are unsaturated polyesters and vinyl esters. However, aircraft applications have more-stringent thermal, mechanical, and environmental performance requirements than can be provided by these systems. Epoxy resins are called for when high-level physical properties are demanded and when the application requires elevated-temperature property retention. Epoxy resins are more expensive, in terms of both material cost and processing. Their reaction rate is significantly slower because they are cured by a stepwise reaction. They also have a short pot life, which translates into a high resin scrap rate. In addition, epoxy pultrusions often exhibit poor surface quality because of a tendency for the resin to adhere to the die wall. A number of investigations have focused on making epoxy systems more processible through viscosity-modifying fillers, internal mold releases, and hybrid epoxy structures (Martin and Sumerak, 1987).

As described earlier in this chapter, interest in continuous-fiber-reinforced thermoplastic-matrix composites has

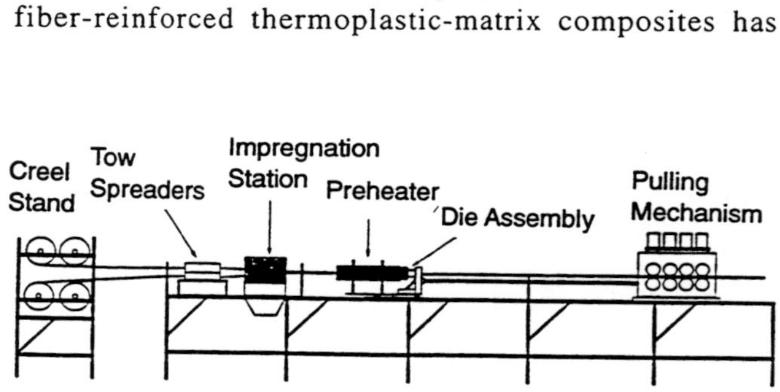

FIGURE 4-5 Composite pultrusion process. With permission from University of Delaware Center for Composite Materials.

recently increased. For pultrusion processing, thermoplastics also offer post-forming capability, superior weatherability, cosmetic advantages, abrasion resistance (Goldsworthy and Martin, 1985), and superior heat distortion properties (Martin and Sumerak, 1987). However, thermoplastic pultrusions have not gained broad market acceptance because of the relatively high, final composite cost compared to the thermoset alternative; the high cost of raw materials for thermoplastic resins and relatively inefficient manufacturing methods are the primary cost drivers.

In thermoplastic pultrusion, the prepregs are guided into a preheater, where the material is heated to a temperature near or in excess of the melting point of the matrix. The material then enters a heated die with a significant taper that gradually shapes the material to the desired cross-section where a cooled die consolidates the newly formed composite by lowering the composite temperature below the melting point of the matrix. The cooled die is followed by a cooling-off section before the composite is clamped and pulled forward by the pulling mechanism and then cut to suitable lengths with a saw.

Virtually all work on thermoplastic pultrusion has used prepregs as raw material, including not only solvent- and melt-impregnated materials but also powder-impregnated and commingled materials. The materials that have been used include all common kinds of reinforcements (glass, carbon, and aramid) and matrices such as polypropylene, polyetherimide, polyetheretherketone, and many others. Processing problems stem largely from the difficulties of dealing with stiff prepregs and the very high viscosities of thermoplastic resins (1,000,000 cp, compared with 500–1000 cp for the thermoset resins used in pultrusion) (Wilson et al., 1989).

Advanced Tow Placement for Thermosets

Automated tow placement using thermosetting materials was pioneered by Hercules in the early 1980s. The process features strict control of ply orientation and thickness with complex geometries required for wing skins. Ply thickness in a part with tapering or changing cross-sections is maintained by incrementally narrowing or widening the band of material as it is laid down through addition or deletion of individual tows[1] of material.

Automated tow placement combines the best features of filament winding, automated tape laying, and multiaxis robotics. Thermoset structures fabricated using this technology achieve properties similar to structures fabricated with prepreg tape. The technology is applicable to a wide product range and has actually been demonstrated on the aft fuselage of the V-22 Osprey, a project that demonstrated the cost benefit of the process—61 percent compared with hand laid-up fabric.

The thermoset tow-placement process typically requires autoclave curing. In thick sections, multiple debulking steps or incremental cures may be required to achieve high quality (e.g., low void content, low-layer waviness). These features of the thermoset process, along with the high cost of processing equipment, contribute significantly to the cost of the structure.

In-Situ Consolidation of Thermoplastics

Automated thermoplastic tow placement is an emerging nonautoclave process with potential for economically fabricating large structures. Unlike thermosetting polymers, which undergo irreversible crosslinking during processing, thermoplastics are simply melted and fused when processed. Thermoplastic composites can be processed at very high throughput by heating locally to the required process temperature and applying local consolidation pressure. With on-line consolidation occurring in thousandths of a second, the technical risk with thermoplastic tow placement is associated primarily with the ability to deliver and control the necessary heat and compaction pressure at the delivery point. Controllable heat sources and compaction methods are required that allow throughput rates which achieve process cost-effectiveness. Further breakthroughs in real-time inspection are key to affordablity by minimizing post-NDE inspection.

Electron-Beam Curing

Electron-beam (e-beam) curing is an emerging manufacturing technology that offers high potential as a low-cost, nonautoclave process for cure of large composite structure. The advantages of electromagnetic processing to cure polymer-matrix composites include the following (Goodman et al., 1996):

- Reduced curing times. Although products are cured individually with e-beam rather than in batches as with autoclave processing, the production time is expected to be reduced.
- Continuous operation. Components can be further processed as soon as they are fabricated, providing simplified production scheduling and inventory control.
- Improved resin stability. Electron-curable resins do not have to be stored at low temperature because they are less reactive at room temperature than formulations designed for thermal curing.
- Increased design flexibility through process control. Because radiation can be controlled, materials with

[1] A tow is a bundle of continuous filaments that makes up a reinforcing fiber.

different thermal curing cycles can be combined in a single product.

While e-beam cure offers great benefits, few fundamental studies have been conducted to understand how the process and the resulting microstructure affects the long-term performance of these materials. Consolidation is needed to remove voids and is difficult in this nonautoclave process. Little is understood about the radiation chemistry of candidate matrix composites. New materials and process development to fully exploit the potential of this technology is required for applications in the next generation of civil aircraft.

SUMMARY

The application of polymeric composites has been an evolutionary process, with increased use as materials and processing technology matured and program needs dictated their use. The factors currently driving new materials applications on commercial aircraft will place added emphasis on design simplification, low-cost processing, and durability and maintainability. The committee believes that continuing developments in toughened thermosetting polymer and high-performance thermoplastics that are compatible with innovative processes are the materials advances that will contribute to these goals. The materials optimization efforts will take a balanced approach that considers both small-scale (including most impact threats) and large-scale (including penetrations and severed elements) damage in establishing performance criteria.

While the development of composites for aerospace applications has traditionally been driven by performance, cost has assumed increasing importance during the past several years. Thus, a primary criterion in the development of manufacturing processes for the next generation of commercial transports has been the potential for low-cost production of components. The committee believes that the trend to develop low-cost production processes such as resin transfer molding, resin film infusion, diaphragm forming, pultrusion, advanced tow placement, and nonautoclave processing will be continued for the foreseeable future.

5

Environmentally Compliant Materials and Processes

Concern for the environment and the health and safety of workers has become a critical criterion for the selection and application of materials and processes in most manufacturing industries. In the aircraft industry, the concerns center around the organic solvents used in many manufacturing processes and materials systems as well as heavy metals such as chromium and cadmium used in corrosion-resistant coatings. Environmental factors are having a profound effect on the design of new aircraft, the production of current models, and the operation and maintenance of the existing fleet.

ENVIRONMENTAL REGULATIONS

Federal health and environmental laws have a direct effect on aircraft materials and processes.[1] Some federal laws, including the Resource Conservation and Recovery Act (RCRA), the Clean Air Act (CAA), and the Clean Water Act (CWA), attempt to decrease waste by restricting releases of certain chemicals or compounds and by raising the cost to industry of releasing wastes into the land, air, and water. The Toxic Substance Control Act (TSCA) directly controls the use of hazardous chemicals.

Resource Conservation and Recovery Act

RCRA involves solid waste management and requires an inventory of releases of certain materials. The most critical materials affected by RCRA are corrosion-inhibiting materials containing heavy metals such as chromium and cadmium. RCRA also places emphasis on reducing waste through recycling or reuse, affecting both in-process waste and end-of-life disposal. The Comprehensive Environmental Response, Compensation and Liability Act, or "Superfund," provides for penalties and assignment of liability for wastes improperly disposed of under RCRA.

[1] An analysis of the effect of environmental policy on product design is presented in *Green Products by Design* (OTA, 1992).

Clean Air Act

The CAA involves the control of ozone-depleting chemicals in industrial processes. It requires the development of state implementation plans to achieve national, ambient air quality standards. This legislation has a profound influence on aircraft design and operation, because it affects all materials and processes involving volatile organics, including paints and finishes, cleaning and surface preparation, and adhesive bonding. The Clean Air Act's National Emission Standards for Hazardous Air Pollutants (NESHAP) extends the Montreal Protocol's solvent elimination restrictions to require the use of lower-vapor-pressure solvents. Also, NESHAP addresses the reduction and replacement of inorganic hazardous air pollutants such as chromium and cadmium.

Clean Water Act

The CWA involves control and inventory of pollutants discharged to navigable waters through the National Pollution Discharge Elimination Program. While there is an effect on manufacturing processes such as cleaning and surface preparation, the greatest effect will be on aircraft operations such as paint removal, cleaning, and de-icing.

Toxic Substance Control Act

TSCA requires manufacturers to obtain approval from the Environmental Protection Agency (EPA) to produce or use new chemicals that may represent an environmental or health risk. TSCA requires the development of materials safety data sheets to document compliance. The EPA may grant limited exemptions for laboratories or facilities involved in the development and testing of new materials that have not yet been fully characterized.

WORKPLACE HEALTH AND SAFETY

New materials and processes must be developed so that they do not create new health or environmental safety problems. The transporting and processing of the materials are coming under more-strict regulation to protect the health of workers involved in production and distribution.

The health and environmental effects of new materials are regulated under the TSCA. The act requires reporting of production levels, use categories, and exposure data from manufacturers and processors of these chemicals.

A complete evaluation of toxicity can take as long as five years and is quite costly. Concern about a substance's health effects may arise if there are indications of carcinogenic effects, acutely toxic effects, serious chronic effects, or adverse environmental effects, or if concern exists about the health or environmental effects of one or more impurities or byproducts of the substance.

All of the health and safety issues have to be considered in relation to the potential exposure of individuals. For example, a new material could be toxic and still be acceptable if exposure can be minimized by taking the appropriate actions during its manufacture and use. Emphasis has been increasingly placed on approaches to consider health and safety concerns in the original materials selection processes to yield a final product that is safer both to the environment and to the individuals who will be exposed. The effective implementation of such an approach requires methods for assessing toxic potential and to predict the effects on workers.

EFFECT ON AIRCRAFT MATERIALS

The effect of new materials or processing technologies on worker health and safety and on the environment have become critical factors in their development and introduction. It is important to consider environmental factors throughout the materials life cycle, including raw material extraction and manufacture, transport and handling, fabrication, utilization, and disposal. For materials suppliers, fabricators, aircraft manufacturers, and the airlines, these concerns represent a significant risk and expense in the development of new materials or processes. Environmental regulatory requirements have caused the aircraft manufacturers, suppliers, and end-item users to undertake significant research and development of environmentally compliant materials and processes for implementation purposes. Significant efforts and resources have been directed toward compliance with a myriad of federal, state, and local environmental regulations. In addition, because of the global nature of the aircraft and airline industries, international environmental regulations can have a profound effect on the manufacture and operation of the aircraft fleet.

Because of the continued concern for the environment, federal (Occupational Safety and Health Administration and EPA), as well as state and local regulations, are becoming increasingly restrictive and may constrain technology advancements until industry learns to further reduce waste, monitor and control pollution, establish tolerable standards, and dispose of or collect hazardous substances. The cost of responding to changing regulations may in some cases be intolerable for some technologies, especially in the initial stages of technology development.

Compliance activities by the aircraft industry include extensive facility modifications; new equipment research and procurement; chemical evaluation and reformulation for solvents, paints, paint strippers, and structural adhesive primers; maintenance procedure revisions; maintenance and repair training; and safety modifications.

Federal Aviation Regulations require the airlines to maintain and repair aircraft using materials and processes that are identical or equivalent to those used in the original aircraft manufacture. However, whether the aircraft is new or old, compliance with environmental regulations is mandatory. Airlines, although assisted by manufacturers' research for process modifications, including surface preparation for aluminum prior to bonding, anodizing (boric-sulfuric- replacing chromic acid), and corrosion prevention coatings, must develop many customized maintenance materials to replace processes that are no longer compliant with environmental regulations.

Some of the developments linked to compliance with environmental regulations include:

- chromium reduction and elimination—chromic-acid anodize replacement with boric-sulfuric-acid anodize;
- Zn-Ni alloy plating or ion-vapor-deposited aluminum coatings as alternative to cadmium plating;
- modification of vapor degreasers for aqueous cleaning systems;
- methyl siloxanes as replacements for chlorofluorocarbons and methyl chloroform in precision cleaning;
- replacement of 1,1,1 trichloroethane in handwipe cleaning operations;
- computerized inventory data management of hazardous materials, material tracking, and management; and
- new coating-removal developments, methylene chloride-free paint strippers, low-volatile organic compound coatings, replacement of primers containing ozone depleting substances.

As restrictions on disposal become more stringent worldwide, materials suppliers and aircraft manufacturers must consider the issues of recycling and disposal of advanced materials. These considerations must include means to recycle, remanufacture, or reuse in-process scrap as well as components that have reached the end of their service life. The environmental impact of new products can be reduced by considering "green" design issues, such as waste prevention (e.g., through improved materials utilization and component durability) and better materials management (e.g., recovery, disassembly, and separation considerations), throughout the materials development and component design processes (OTA, 1992).

SUMMARY

Concerns related to environmental compatibility and the health and safety of workers have a significant influence on the development and application of all new materials and processes. It is important to consider environmental factors throughout the entire materials life cycle, including new materials extraction and manufacture, transport and handling, fabrication, utilization, and disposal. For next-generation commercial transports, the effects are particularly significant in materials that contain heavy metals or volatile organic compounds or in processes that use solvents. In response to regulatory pressures, the drive to new materials and processes will have the greatest impact on the operation and maintenance of the aircraft fleet in the areas of corrosion protection, surface preparation processes, finish materials, and finish application processes. The FAA should assess the effect of these material changes on the aging of the new aircraft fleet.

III

Analytical Methods

6

Methodologies for Assessment of Structural Performance

Current structural design and analytical procedures used by the aeronautics industry are largely semiempirical, even though significant improvements have occurred in structural analysis methodology over the last two decades. Preliminary design efforts require estimates using relatively quick, easy to use, and insightful tools to allow sensitivity studies among different structural design options. Final design efforts require tools that provide a precise and accurate assessment of the structural design.

Finite element analysis methods are routinely used for predicting the stress, strain, and displacement fields in complex structural geometries. Superior graphical interfaces have significantly improved pre- and post-processing of data files. Automated mesh generation, mesh refinement, and automated adaptive remeshing have resulted in major improvements in the efficiencies of model development and analysis and in the accuracies of the numerical solutions. Post-processing algorithms and graphical interfaces have significantly enhanced the ability of the analyst to interpret the results of the stress analysis. Along with these improved analytical and software tools, advances in the available computing capabilities have been rapid. In spite of these advances, the reliable prediction of structural failure modes, ultimate strength, residual strength, and fatigue life has remained elusive to the structural engineer. Standard practice still relies heavily on extensive testing at the subelement, element, subcomponent, component, and full-scale levels. Design details are frequently optimized through test programs. Scale-up effects are handled through a building-block approach that relies on testing to verify the anticipated structural performance at each scale level. Full-scale static and fatigue tests are conducted to identify "hot spots," to verify adequate structural integrity for design limit and ultimate loads, and to verify durability and damage tolerance requirements. Hence the designation "semiempirical" for current practice, which is necessarily expensive and time-consuming.

In spite of over 60 years of experience by researchers and industry in designing metallic structure and 25 years of experience designing composite structure, the lack of rigorous analytical methods to predict residual strength and fatigue life can inhibit the cost-effective introduction of advanced materials with superior specific strength and stiffness relative to current conventional materials. Continuing efforts are needed to develop—and insert into standard engineering practice—advanced mechanics-based analytical prediction methodologies that would allow innovative designs to be evaluated and optimized at acceptable cost. Such methods would also provide the means to assess the effects of service history on the durability and damage tolerance performance of the structure. This chapter examines current issues and assess the impact of advanced methods on the introduction of new materials and the design of more-structurally efficient and cost-effective primary structures on next-generation aircraft.

ANALYTICAL PREDICTION METHODOLOGY FOR STRUCTURAL INTEGRITY

Metallic Materials

Metallic materials tend to fail due to the formation and growth of a dominant microcrack that eventually reaches a critical length and then more rapidly propagates to failure. However, the recently recognized phenomenon of multiple-site damage has been shown to be a critical issue in aging of commercial transport aircraft. While fracture mechanics is now a mature part of standard practice in engineering, rigorous prediction methodology only exists for brittle materials that exhibit limited plasticity. The fatigue crack growth behavior and fracture processes exhibited by ductile materials are reasonably well understood. However, the development of rigorous elastic-plastic analytical methods has been hampered by the complicated three-dimensional effects present in most structures. In thin-sheet planar structural components, local constraint effects on the development of plastically deformed material frequently result in highly inaccurate solutions when obtained from two-dimensional plane-stress or plane-strain assumptions. Three-dimensional analyses of cracks in geometries such as lugs and fittings may be inaccurate due to uncertainties in modeling the crack-front singularity.

Three-dimensional models of the geometry and crack configurations are computationally intensive and have been impractical for the practicing engineer. Therefore, improved engineering methods for fracture mechanics using two-dimensional approximations that yield reliable results are required for more-accurate, structural integrity prediction methodology. In addition, methods that compute reliable, stress intensity factors for various crack geometries using stresses obtained from uncracked stress analysis models should also be developed.

Most fracture mechanics analyses are performed on refined models of local structural details that use stress boundary conditions determined from a global structural analysis. Therefore, the computed stress intensity factors will never be more accurate than the global structural analysis, regardless of the sophistication of the fracture mechanics model. The structural models must account for global details such as frames, stiffeners, tear straps, shear ties, and shear clips; must accurately capture local details such as interference-fit stresses and clamp-up stresses at riveted fasteners; and must also accurately treat combined load effects and geometric nonlinear behavior. This level of fidelity in the local stress analysis can only be achieved by the development of more-robust and computationally efficient structural analysis methods that exploit global-local and other hierarchical computational strategies. However, it is not sufficient to only generate highly accurate local stresses. Other features that promote crack initiation and growth need to be modeled as well. These features include metal-forming defects, machining and fabrication defects, residual stresses such as from drilling holes and forming rivets, and mechanisms such as fretting and pitting corrosion. These features may be modeled best by an approach such as the "equivalent initial flaw size" to facilitate the fatigue and residual strength analysis.

The design of longer-life and more-durable metallic structures could also be significantly enhanced by the incorporation of models that reliably account for the effects of the environment on fatigue life and residual strength. The aluminum alloys currently used in airframe structures are highly susceptible to corrosion damage. Elaborate corrosion prevention and control programs are used by the manufacturers and airline operators. These programs are successful at delaying the development of widespread corrosion, but are not 100 percent effective in preventing localized effects such as pitting that may cause crack nucleation. Furthermore, fatigue crack growth rates may be significantly accelerated when newly created fracture surfaces are exposed to the environment.

An improved understanding of the role of the environment on alloy behavior and better analytical methods to reliably predict spectrum loading effects, such as load sequencing, overloads, and stress reversals, on fatigue life will remove conservatism in current designs and may accelerate the use of advanced materials for corrosion-critical components. For those alloys that exhibit a strong crack-closure effect, models should be developed that incorporate the effects of the environment into closure mechanics such as plasticity-induced closure. For other materials, superposition models need to be developed which attempt to treat time-dependent effects and cyclic mechanical load effects independently. Typically, these models are computationally simple and rely on Paris Law formulations, but may require more test data to fully characterize mean stress and stress ratio effects.

The effects of widespread fatigue damage, including multiple-site damage (MSD) and multiple-element damage (MED), have been critical in the understanding of aging of metallic aircraft structures. Structural mechanics research is needed to develop analytical methods to predict the residual strength of a thin-sheet stiffened shell structure with fatigue cracks or damage caused by an in-flight accident. Structural analysis methods are needed to evaluate the damage tolerance of aircraft for MSD and MED and to provide a basis for assessing the integrity of repairs on existing aircraft structures. Stiffened shell analysis methods should include automated, adaptive remeshing capability, as well as a capability to analyze structures with nonlinear materials properties and at large deformations. These improved methods should be integrated into the structural analysis methodology to predict fatigue crack growth and residual strength of aircraft structures. In cooperation with industry, a structural test program, through the subcomponent and subscale test article level, will be necessary to verify the integrated fracture mechanics and structural analysis methods.

Composite Materials

In contrast to metallic alloys, composite materials exhibit complex failure modes that may involve the interaction of several different damage mechanisms. The complexity of the failure modes becomes a major design consideration when addressing durability and damage tolerance requirements. The prediction of composite structural behavior is complicated by the fact that there are virtually no material properties of the constituents that can be measured independently and used to predict the response of the composite. Most design allowables depend on the as-cured properties of the constituents and the fiber architecture. The process of inferring local (micro) behavior from global (macro) response tends to promote empirically based design practices.

The current design practices rely heavily on empirical approaches to conservatively estimate the effects of damage on durability and damage tolerance. For example, simple tests such as open-hole compression and compression-after-impact tests are often used to establish "knock-down" factors for reducing the working strain levels to avoid damage growth. In the absence of mechanical methods to predict damage initiation, growth, and failure modes, all test data must be obtained on the specific laminate stacking sequence or fiber preform architecture to be used in the design. As described in the introduction to part 2, structural concepts are tested at the element and subcomponent level to confirm the suitability of the structural design parameters and to couple coupon-level material behavior and test data to the actual structural behavior. This approach would be significantly enhanced by the introduction of rigorous, mechanics-based analytical methodology that would enable the design of composite structures that are more structurally efficient.

There are several key technical barriers to developing reliable methods to predict failure of composite structure. The first barrier is related to the complex three-dimensional stress states produced by the fiber architecture. The task of developing crack initiation and growth criteria for complex local effects is complicated by the difficulty of isolating an individual damage mechanism and varying the stress states to develop a comprehensive failure criterion.

Second, loading-history-dependent growth laws for individual damage mechanisms and for interacting cracks do not exist. Progress has been made developing damage onset criteria, particularly for those special cases where an individual damage mechanism can be isolated, but three-dimensional models of the material system may be required to predict the correct local stresses to use in the criteria. This results in a massive computational process involving an iterative and incremental loading scheme and may require a very refined finite element mesh with many degrees of freedom. Also, neither fatigue crack growth and life nor stable damage progression prior to catastrophic fracture can be predicted in a rigorous sense because crack growth laws do not exist. Once again, empirical engineering approaches have been developed to avoid these complexities.

Third, residual stresses in the matrix polymer that result from elevated-temperature processing (i.e., curing of thermosets or consolidation of thermoplastics) need to be characterized. Processing residual stresses often leads to the first damage (matrix cracking) in the composite structure. Matrix cracking can eventually lead to delamination, fiber damage, and other damage-mode interactions.

Finally, the understanding of failure mechanisms and associated failure criteria, including interactions that take place between different failure modes at length scales ranging from constituent-level (e.g., matrix cracking) to component-level (e.g., buckling) is incomplete. The complexity of the (inhomogeneous) composite materials and their array of anisotropic material strengths give rise to the development of a corresponding array of damage and failure modes in these materials that must be understood and correctly modeled. Hence, there is a need to develop understandings and representations of the critical damage and failure modes that control the types of structures typical of commercial aircraft.

A report of the National Research Council discusses some of the methods for the assessment and prediction of durability and damage tolerance of composite materials (NRC, 1991). General features of these methods include:

- Remaining strength and life models are developed and predictions are made for each independent failure mode (such as fiber failure in tension or microbuckling in compression).
- Mechanical representations of the state of stress and state of material are constructed on the basis of a "representative volume" of the material that is typical of the distributed damage state that controls the remaining stiffness and strength of the composite.
- Various methods are used to characterize and monitor the degradation rate of the strength of composites.
- Micromechanics (mechanics analysis at the fiber/matrix level of representation) is increasingly used for remaining strength modeling, for the calculation of stiffness change (which leads to internal stress redistribution), and for the estimation of the remaining strength for a given failure mode.
- Statistical considerations are essential for the correct representation of the long-term behavior of composites. Composites typically fail because of the statistical accumulation of defects, which eventually interact to create a critical condition.
- Time-dependent behavior such as viscoelastic creep, creep rupture (driven by, for example, internal stress redistribution or oxidation), and aging are typically important in the consideration of the long-term durability and damage tolerance of polymer-matrix composites.

Two methodologies for predicting remaining strength (damage tolerance) and life of composite structure for aeronautical applicators are currently being researched and developed for potential use in design. The first is the use of "damage mechanics" to predict the changes in stiffness that occur during service life (Talreja, 1985; Simo and Ju, 1987; Lee et al., 1989; Shapery, 1990). This approach is becoming common for the purpose of following the development of damage and for interpreting the changes in stiffness of the structure as well as at the microstructural level in composites.

The second methodology is the use of micromechanics and kinetic theory to predict remaining strength. A schematic diagram of such an approach appears in figure 6-1. Micromechanical representations of the fundamental composite strengths are constructed in terms of the constituents, their geometry, and their arrangement (Reifsnider and Stinchcomb, 1986; Reifsnider, 1991a,b, 1992; Reifsnider and Gao, 1991; Gao and Reifsnider, 1993; Xu and Reifsnider, 1993). The constitutive parameters in these strength models are studied as a function of the service inputs and environments using kinetic (or rate) theory, allowing fatigue, creep, creep rupture, aging, oxidation, and other time-dependent and cycle-dependent effects to be introduced at the constituent level. These constituent effects are combined and their collective effect assessed by the micromechanics models mentioned

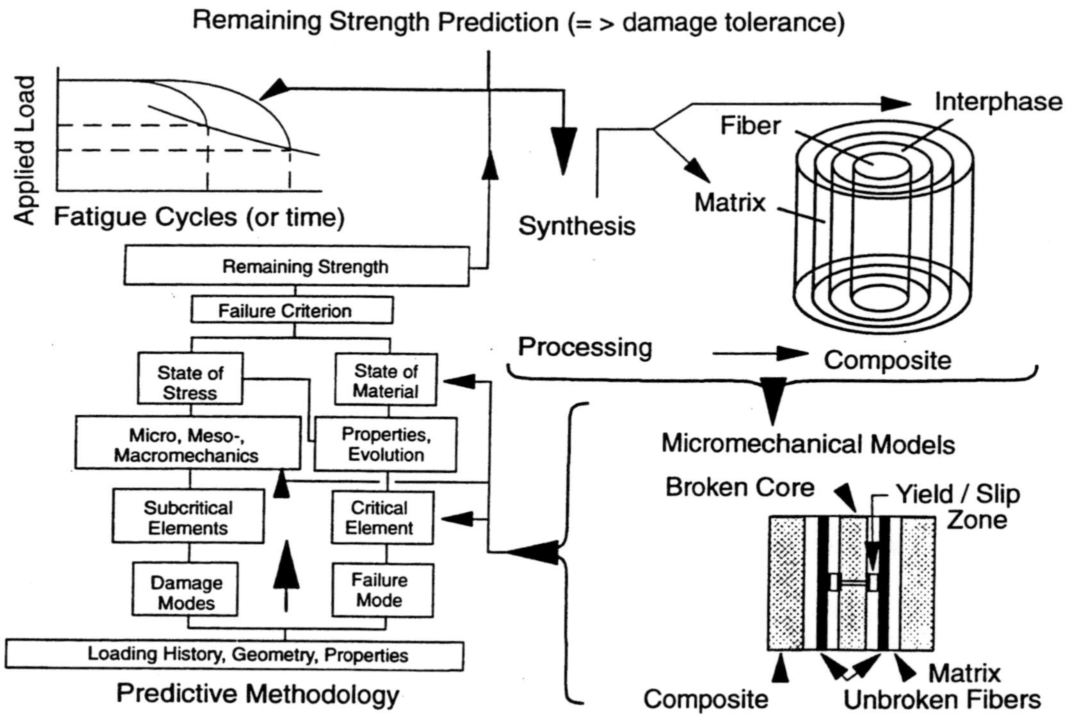

FIGURE 6-1 Methodology for predicting remaining strength in composites.

earlier, without the need for empirical or phenomenological postulates, which is a major advantage.

More reliable and robust mechanics-based analytical methodology to predict structural failure modes, life, and residual strength need to be developed to support the introduction of new materials and to contribute to the development of efficient and optimized structural designs. While semiempirical engineering approaches will continue to be refined, the most promising methods are those that will exploit the dramatic improvements in computer capability and numerical methods. The three-dimensional behavior of composite materials is well suited to global, local, and hierarchical methods of analysis. New methods can efficiently couple local material-level damage mechanisms to global responses, such as load redistribution, geometric nonlinear and instability phenomena, and predict structural failure modes.

Adhesive-Bonded Joints

Key structural integrity issues continue to limit the widespread acceptance of adhesive bonding as a viable joining technology for commercial transport aircraft. This is partly because of negative experiences with early bonding technology and partly because of unresolved technological barriers such as those described in chapter 2 (e.g., inspection methods). Both strength-based (Hart-Smith, 1990) and fracture mechanics-based (Rakestraw et al., 1994) design methods have been developed to determine appropriate bonding parameters to achieve the desired load transfer capability.

Interfacial properties are crucial in the understanding of long-term durability of adhesive-bonded joints, especially under harsh environmental conditions encountered in aircraft service (Kinloch, 1979; Kinloch and Osiyemi, 1993). Fracture mechanics testing and analytical techniques can be used to provide quantitative evaluations of adhesive-bonded structure (Rakestraw et al., 1994). Further work is needed to develop fracture mechanics-based adhesive tests, such as double-cantilever-beam and edge-notch bend tests, and analytical techniques that can provide information concerning rate-dependent failure characteristics and the influence of environmental effects.

Surface preparation and bonding processes are very unforgiving, placing more reliance on inspection technology to assess structural integrity. While there are methods under development that appear to have high reliability to detect disbonded regions, there are currently no reliable methods to assess bond integrity in cases where there are no disbonds but the existing bond is of poor quality and does not have the expected load transfer and strength characteristics. This basic problem must be resolved through improved bonding technology, better fail-safe designs, or more reliable inspection methods before expanded applications of adhesive bonding will be implemented into new aircraft designs.

ADVANCED COST AND STRUCTURAL OPTIMIZATION METHODOLOGY

Major advances in computing technology can have a significant impact on the cost-effective incorporation of high-performance materials into both new and derivative aircraft. At this writing, optimistic projections suggest the availability in several years of moderately priced, powerful desktop computers with massively parallel processing (MPP) capabilities. This computing power would enable designers and analysts to make rapid, theoretical performance analyses of a very large number of candidate material and structural configurations. Obviously, the results would be no more reliable than the assumptions underlying the numerical formulations. However, in conjunction with the development of the improved mechanics-based analytical methodologies described in the previous section, the computing power afforded by MPP should go a long way toward reducing the time-consuming, expensive testing now needed to validate the performance of new materials and structural configurations.

Cost-effectiveness has already been emphasized as a preeminent driver in the present industrial climate facing the airlines. But estimating the real cost of introducing a new material into an aircraft is a task of incredible difficulty, requiring assessments of costs of myriads of contributory elements—not just the obvious cost of acquisition, processing, manufacturing, engineering, maintenance, but also the elusive costs of money, time, and risk. (In comparison, stress analysis is trivial!) A willingness by industry, government, and academia to grapple with sophisticated, rational cost analyses is starting to emerge. This trend should be encouraged. With precise, visible exposure of assumptions and definitions, and prudent recognition of uncertainties, credible cost analyses can help industry steer a sensible course between competition-vulnerable stagnation and risky, premature innovation. Furthermore, the full potential of such cost modeling and analyses dictates its concurrent integration into the high-speed, high-power, numerical design and analysis procedures anticipated in the future. The great danger, to be assiduously guarded against, is that cost models will start to be treated as natural laws. They will never be exactly right; by their nature they will need unending revision. With appropriately chosen criteria of merit, an integrated design-analysis-cost model can provide the basis for a numerical optimization package that, in conjunction with MPP, would constitute a powerful tool for the cost-effective utilization of new materials.

SUMMARY

Current structural design and analytical procedures used by the aeronautics industry are largely semiempirical, even though significant improvements have occurred in structural analysis methodology over the last two decades. Continuing efforts are needed to develop—and insert into standard engineering practice—advanced mechanics-based analytical prediction methodologies that would allow innovative designs to be evaluated and optimized at acceptable cost. Such methods would also provide the means to assess the effects of service history on the durability and damage tolerance performance of the structure.

For metallic structures, fracture mechanics-based methods are fairly well developed. Work is needed to improve the analytical capabilities that account for the effects of crack and component geometry; the effects of stress concentrations and manufacturing defects, corrosion, environmental exposure; and existing fatigue damage. Improved methods should be integrated into the structural analysis methodology to predict fatigue crack growth and residual strength of aircraft structures.

For composite structure, complex failure modes that may involve the interaction of several different damage mechanisms are major design considerations when addressing durability and damage tolerance requirements. There are several key technical barriers to developing reliable methods to predict failure of composite structure, including incomplete understanding of complex three-dimensional stress states, damage progression laws, processing residual stresses, and understanding of failure mechanisms and associated failure criteria.

For adhesive-bonded structure, further work is needed to develop fracture mechanics-based adhesive tests and analytical techniques that can provide information concerning rate-dependent failure characteristics and the influence of environmental effects.

Cost has been identified as an important driver in the application of new materials and structures technology on commercial transports. With appropriately chosen criteria of merit, an integrated design-analysis-cost model can provide the basis for a numerical optimization package that would constitute a powerful tool for the cost-effective utilization of new materials.

IV

Aircraft Operations

Consideration of aircraft operations, including inspection, maintenance, and repair procedures is crucial in the development and application of new materials and structures. This part of the committee's report focuses on the operation and monitoring of materials and structures in a service environment.

This part is organized in two chapters:

- Chapter 7, "Aircraft Maintenance and Repair," describes the issues related to maintenance of commercial transport aircraft. The lessons learned from the aging of metal and composite structure are discussed.
- Chapter 8, "Nondestructive Evaluation," describes current aircraft inspection practices and identifies needs for improved nondestructive evaluation techniques and promising technologies for the future.

7

Aircraft Maintenance and Repair

The successful utilization of new materials and structural concepts relies on maintenance programs that cost-effectively ensure passenger safety. This chapter is an overview of the current experience in aircraft maintenance programs, including inspection and repair processes, lessons learned from aging aircraft, and future needs to support new materials and structural concepts.

AIRLINE MAINTENANCE

Maintenance programs are evolved and developed for each new type of aircraft based on previous experience with similar materials, engines, components, or structures. New materials or structures, for which experience is limited, are observed more frequently until a basic level of confidence is established. Time extensions to inspection intervals are based on observations made during routine service checks. A typical airline maintenance and service plan is outlined in table 7-1. The objectives of an effective maintenance program are as follows (Edwards, 1994):

- Ensure, through maintenance activity, that the inherent safety and reliability imparted to an aircraft by its design are sustained.
- Provide opportunities to restore levels of safety and reliability when deterioration occurs.
- Obtain information for design modification when inherent reliability is not adequate.
- Accomplish the above at the lowest possible cost.

Structural Maintenance

Any new aircraft program is based on assessing structural design information, fatigue and damage tolerance evaluations, service experience with similar aircraft structures, and pertinent test results. Generally, the maintenance task evaluates sources of structural deterioration including accidental damage, environmental deterioration, and fatigue damage; susceptibility of the structure to each source of deterioration; the consequences of structural deterioration to continuing airworthiness including effect on aircraft (e.g., loss of function and reduction of residual strength, multiple-site or multiple-element fatigue damage, the effect on aircraft flight or response characteristics caused by the interaction of structural damage or failure with systems or power plant items, or in-flight loss of structural items); and the applicability and effectiveness of various methods of detecting structural deterioration, taking into account inspection thresholds and repeat intervals.

Component Maintenance

The application of new materials will not cause undue maintenance difficulties or hardship for the airlines provided the aircraft designer is familiar with component experience. Airline experience indicates that hardware items wear out, but statistical old-age wear-out in complex mechanical, electrical, and avionic components is not a dominant pattern of failure. In fact, over 90 percent of generic part types show either random distribution of failure or gradually increasing probability of failure with age (Edwards, 1994).

The reliability of a part or component of aircraft hardware is only as good as its inherent design (supported by adequate maintenance) allows it to be. Hence, it is generally accepted that (1) good maintenance allows parts to reach their potential reliability; (2) overmaintaining does not improve reliability, but does waste money; and (3) undermaintaining can degrade reliability. In general, fundamental design changes are required to correct inherent component reliability problems.

There are three approaches to preventative maintenance that have proven to be effective. The first method, *hard time*, involves removing a unit from service when it reaches a pre-ordained parameter value. The second method, *functional check* or *inspection*, involves monitoring a characteristic dimension or usage/operating parameter of a piece of hardware to determine if it is still suitable for continued operation, or if it should be removed to prevent an in-service failure. The third method, *functional verification*, requires performing an operational check of hardware function(s) to determine each function's availability if it is normally hidden from the scrutiny of the flight and operating crew.

There are many components for which measurement of deterioration, periodic removal for maintenance, and hidden function verification are not economically feasible or beneficial. Such parts require routine performance or reliability

TABLE 7-1 Typical Airline Maintenance and Service Plan

When Service is Performed	Type of Service Performed	Impact on Airline Service
Prior to each flight	"Walk-around"—visual check of aircraft exterior and engines for damage, leakage, and brake and tire wear	None
Every 45 hours (domestic) or 65 hours (international) flight time	Specific checks on engine oils, hydraulics, oxygen, and specified unique aircraft requirements	Overnight layover service
Every 200–450 hours (22–37 days) flight time	"A" check—detailed check of aircraft and engine interior, services and lubrication of systems such as ignition, generators, cabin, air conditioning, hydraulics, structure, and landing gear	Overnight layover service
Every 400–900 hours (45–75 days) flight time	"B" check (or "L" check)—torque tests, internal checks, and flight controls	Overnight layover service
Every 13–15 months	"C" check—detailed inspection and repair program on aircraft engines and systems	Out of service for 3–5 days
Every 2 years (narrow-body aircraft)	Inspection and reapplication of corrosion protective coatings	Out of service up to 30 days
Every 3–5 years	Major structural inspections with attention to fatigue damage, corrosion, etc. Aircraft is dismantled, repaired, and rebuilt. Aircraft is repainted as needed	Out of service up to 30 days

monitoring, and no preventive maintenance is required or desirable. Modern aircraft are more tolerant of failures than older aircraft designs because of the increased redundancy provided in the design.

Generally, most airlines classify specific component maintenance tasks as follows:

- lubrication or servicing, where the replenishment of the consumable reduces the rate of functional deterioration;
- operational or visual check, where identification of the failure must be possible;
- inspection or function check, where reduced resistance to failure must be detectable and the rate of reduction in failure resistance must be predictable;
- restoration, where the item must show functional degradation characteristics at an identifiable age, have a large proportion of units survive to that age, and be able to be restored to a specific standard of failure resistance; and
- discard, where the item must show functional degradation characteristics at an identifiable age, and a large proportion of units are expected to survive to that age.

Malfunctions of components should be evident to the operating crew, have no direct adverse effect on safety (whether they occur as a single or multiple event), and minimize the effect on the operation of the aircraft itself.

SERVICE EXPERIENCE

Effective application of new materials on commercial aircraft requires the designer to consider potential sources of damage or degradation in operating environments and to develop a maintenance and repair approach to address them. Damage may occur due to flight loads, thermal and environmental cycles, and aircraft operation and servicing activities. A number of valuable lessons have been learned from previous experience with metallic and composite structure in the current fleet. These lessons provide evaluation criteria in the application and servicing of new materials and structures.

Ramp and Maintenance Damage

An International Air Transport Association survey estimates that 36–40 percent of damage to aircraft is from ramp and maintenance damage, sometimes called friendly foreign object damage (IATA, 1991). Figure 7-1 shows a diagram of the Boeing 777 aircraft interfaces with servicing and other equipment (Boeing, 1994b). These areas are especially prone to damage and require robust material performance in these locations.

To determine the extent of groundhandling damage, 11 airline operators were queried for ground damage history during the years 1990 to 1993 (Boeing, 1994a). Of the 2,241 incidents reported, more than a third were from unknown causes. A tabulation of the causes of damage is given in table 7-2.

Ramp and maintenance damage can represent significant costs to the airlines. The repair of a damaged component is only part of the cost. The airline also bears the cost of flight delay or cancellation and the effects on connections and aircraft rotations.

TABLE 7-2 Causes of Ground Damage to Aircraft

Cause of Failure	Number of Incidences
Unknown	773
Catering	137
Belt loader	122
Loader	101
Lavatory and water service	66
Container	44
Jetway	246
Baggage cart	127
Tug/towbar/taxi	109
Maintenance	86
Cargo loading	50
Fueling	33

SOURCE: Boeing (1994a).

Aging Aircraft

In April 1988, an Aloha Airlines Boeing 737-200 experienced an in-flight structural failure in which a large section of the upper fuselage ripped open and separated from the aircraft. The failure resulted from multiple-site damage (MSD) and corrosion. In this case, MSD was the link-up of

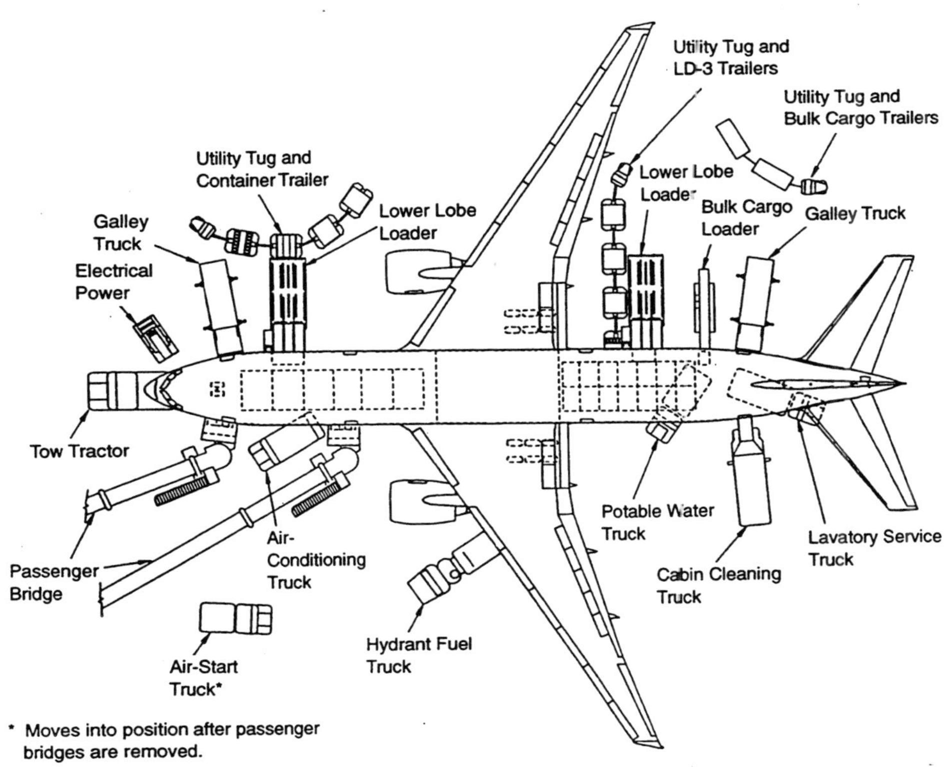

FIGURE 7-1 Diagram of aircraft interfaces with servicing and other equipment. Source: Boeing (1994b).

small fatigue cracks extending from adjacent rivet holes in a longitudinal lap joint in the fuselage. The accident focused international attention on the problems of operating an aging commercial fleet.

In 1990 approximately 46 percent of the U.S. commercial air transport fleet was over 15 years old, and 26 percent was over 20 years old. If current usage and replacement trends continue, the number of aircraft over 20 years old will double by the year 2000. Currently some 3,200 aircraft are affected by FAA Airworthiness Directives that concern operation and maintenance of the aging fleet. The review of experience with aging aircraft has caused an increase in the emphasis on stress corrosion, corrosion, fatigue, and MSD issues. This experience has caused, in turn, the selection of new aircraft alloys with better constituent chemistry control or changes in heat treatment tempers. Also, it has stimulated the development of new organic finishes that significantly retard corrosion, as well as the implementation of design practices to vastly improve corrosion resistance.

The FAA and NASA developed a cooperative research effort aimed at providing a technological basis for ensuring the continued safe operation of the aging commercial aircraft fleet. Each agency developed a program consistent with its mission. The FAA's National Aging Aircraft Research Program addresses the aging aircraft structural safety concerns and provides certification authorities and operators with the tools to meet those concerns. NASA's Airframe Structural Integrity Program is focused on developing advanced integrated technologies to economically inspect for damage and to analytically predict the residual strength of older airplanes. Together these programs form the technological basis for a cooperative effort with U.S. industry to address the critical aging aircraft issues.

Multiple-Site Damage

MSD is a form of widespread fatigue damage that is characterized by small cracks emanating from structural details such as fastener holes (Sampath, 1993). If cracks emanate from adjacent fastener holes, they have the potential to link up and lead to unexpected catastrophic failures as described in the previous section. Also, even without link-up, multiple-site cracks can severely degrade the capability of the structure to withstand major damage from other discrete sources as is described later in this section.

In the past, the standard industry practice was to visually inspect the airframe for damage. Various levels of inspections ranging from daily walk-around inspections to detailed teardown inspections were performed. Instrumented nondestructive evaluation (NDE) methods such as eddy current probes were used only to inspect local regions of the structure where previous cracking problems had occurred. While these inspection methods were labor intensive and highly subjective, they were acceptable because the airframe was designed to survive a two-bay skin crack with a severed frame or stiffener. This design criterion was established to enable the airplane to tolerate major discrete source damage (i.e., such as might be encountered as a result of an engine structural failure) as well as large cracks resulting from the link up of smaller fatigue cracks or the unstable propagation of manufacturing flaws or other service-induced damage. Such damage is large enough that it should be easily detected, and the operator does not need to search for small cracks to ensure the structural integrity of the airframe. However, this "fail-safe" philosophy assumed that the structure adjacent to the major damage (e.g., the two-bay crack) was free of MSD. Design residual strength requirements were based on this assumption. However, the existence of very small cracks (e.g., a few hundredths of an inch or tenths of a millimeter in length) in the adjacent structure can severely degrade this residual strength and thus jeopardize the safety of the airplane as it did in the Aloha Airlines incident. Therefore, inspection of aging aircraft has become much more onerous than for newer aircraft because safety is vitally dependent on the detection of the very small cracks associated with this onset of MSD. This represents a major challenge to the inspection and aircraft industries.

The principal technical needs are (1) to develop and verify advanced NDE technology that can reliably and economically detect disbonds, small MSD fatigue cracks, and corrosion and characterize their effect on the residual strength; and (2) to develop and verify advanced fracture mechanics and structural analysis methodology to predict fatigue crack growth and residual strength of airframe structures to determine in-service inspection thresholds and repeat intervals, quantitatively evaluate inspection findings, and design and certify structural repairs. NDE methods related to MSD are described in chapter 8, and fracture mechanics and structural analysis methods are described in chapter 6.

Corrosion

Corrosion of aging aircraft has been described as an insidious problem (Marceau, 1989). While other aging mechanisms, such as wear and fatigue, are somewhat predictable and can be addressed by the airline maintenance programs to preclude major structural problems, corrosion—especially in its localized forms—is very difficult to predict and detect. Factors that influence the extent of corrosion on aircraft are materials selection, design, component processing and finishing, operational environments, and maintenance programs.

It is anticipated that airplanes manufactured today will experience fewer corrosion problems than those in the current aged fleet because of significant design and corrosion protection improvements that have been implemented and because of operators' increased awareness of the role of these improvements in preventive maintenance. Clearly, maintenance and corrosion control programs will continue to play a major role in the control of corrosion as airplanes age.

Some examples of design improvements to reduce corrosion on the Boeing 777 (Marceau, 1994) include:

- enhanced drainage, especially in the keel of the aircraft;
- the sealing of faying surfaces in corrosion-prone areas;
- the application of improved finish systems;
- liberal use of corrosion-preventive compounds;
- implementation of a good corrosion control maintenance program; and
- improved access for inspection of corrosion-prone areas.

Major airline fleets include aircraft ranging in age from new to 25 years old. Consequently, the degree of corrosion protection incorporated into the airplane varies from limited protection for older aircraft to fairly extensive protection for newer aircraft. Corrosion control programs are tailored to individual fleets, depending on age, prior experience, flight environment and degrees of corrosion protection incorporated prior to the delivery of the aircraft (DeRosa, 1995). All protective finishes are maintained and corrosion prevention compounds are applied during periodic maintenance. Critical areas that are prone to excessive corrosion include areas below the galleys, doorways, lavatories, cargo compartment subfloors, inside external fairings, and the bilges which are all treated at four-year intervals. Landing gear wheel wells and wing spars are treated yearly. Longer intervals of time are allowed between reapplications of corrosion prevention compounds in the case of less-severe environments.

Aging aircraft repairs have typically involved upper-skin lap fastener replacement, nonbonded skin panel replacement, skin lap doubler repairs, frame reinforcement, entryway door and scuff-plate doublers, replacement bushings and clevis joints, bulkhead forging replacement, and selected landing gear component replacement. Based on service experience, the airlines have expectations that manufacturers of new aircraft will (DeRosa, 1995):

- include stress corrosion prevention in all design reviews with airline customers,
- assemble all structure below the aircraft floor with sealant on all faying surfaces,
- coat all detail parts with corrosion-inhibiting primer or polyurethane topcoat before assembly,
- not use adhesive-bonded fuselage skin panels below the floor line or in areas subject to severe corrosion environments such as galleys and lavatories,
- treat all basic fuselage structure with corrosion-preventive compounds, and
- perform a complete (100 percent) inspection for delamination of bonded skin panels prior to the aircraft delivery in order to establish a baseline for subsequent inspection.

The objective of aging aircraft programs is to ensure the continued airworthiness of large transport aircraft as long as they remain in commercial service (Curtis and Lewis, 1992). Because new materials and fabrication processes may yield different degradation and damage mechanisms, a preproduction review should ensure that the new aircraft design includes lessons learned from the existing aging fleet.

Many of the steps needed to improve aging performance are detailed below. Most of these steps have now been incorporated into recent aircraft designs. The susceptibility of aircraft to corrosion and MSD fatigue can be reduced by the following steps:

- eliminating cold-bond lap-joint design details;
- providing adequate drainage to eliminate corrosion in areas where moisture accumulates;
- using the most corrosion-resistant materials and tempers available;
- evaluating galvanic couples with typical coating damage;
- testing dissimilar materials design details for size effects in areas with joints;
- controlling design stress levels, improving design details, and utilizing improved manufacturing and maintenance procedures to preclude the onset of multiple-site damage within the operational lifetime of the airplane;
- developing predictive and monitoring techniques for the onset of MSD (simple and cost-effective techniques should be integrated into the aircraft maintenance plan); and
- providing a complete corrosion prevention and control program within the aircraft maintenance program upon delivery of an aircraft.

The present focus on aging aircraft will lead to better corrosion-resistant treatments for next-generation aircraft. Materials selection in wet areas, the design drainage schemes, the use of insulation standoffs, and sealing and finishing systems have all been improved. The benefits of these improvements should be evident during in-service performance of the Boeing 777 and future aircraft. Liberal use of corrosion-preventive compounds applied in the aircraft assembly process and periodically in service, using a good corrosion control maintenance program, should minimize future corrosion concerns.

Structural Composites

As discussed in chapter 4, prior to the latest generation of aircraft, which includes the Airbus A320 and the Boeing 777, structural composites have been used on aircraft flight control surfaces such as elevators, spoilers, ailerons, and rudders, as well as fairing and fillet panels, landing gear doors, engine cowl doors, and other secondary structures. For these applications, honeycomb sandwich designs with thin 0.6–1.5 mm (0.024–0.060 in.) composite facesheets are most common. It follows that most of the experience with advanced composites has been obtained with this kind of construction. Previously, similar constructions with fiberglass skins and nonmetallic honeycomb core have been used. There is much less service experience with thicker-skin laminate designs that have been used in composite primary structure.

In general, the service experience with composites indicates that damage occurs because of discrete sources such as impacts, lightning strikes, and handling rather than progressive growth caused by a fatigue condition (Blohm, 1994). In addition to groundhandling damage, a recent survey by the International Air Transport Association, summarized in table 7-3, lists the particular causes of damage that occur in the current generations of composite structure (IATA, 1991).

TABLE 7-3 Most Common Causes of Composite Structure Damage to Aircraft

Cause of Failure	Percent of Incidences
Moisture and chemical fluids attack	30
Other (heat damage, fatigue, abrasion, and erosion)	11
Bird strikes and hail damage	8
Runway rocks and foreign object damage	8
Lightning strikes	7

SOURCE: IATA (1991).

The types of damage to composite components include disbonds or delaminations (45 percent), holes or punctures (35 percent), cracks (10 percent), and other damage (10 percent). An especially difficult maintenance issue resulting from these types of damage is when perforation allows the incursion of hydraulic fluids, water, and other liquids into the honeycomb core. Composites may also suffer loss of load-bearing capability due to resin charring and the potential for corrosion of adjacent metallic surfaces. Typical causes of composite service damage mechanisms are shown in table 7-4.

Service experience with thicker composite laminate constructions, such as that used on primary structures on the Airbus A320 and Boeing 777, is not adequate enough to establish damage trends.

Composite Repair

The current methods used by the airlines to repair damage to aircraft composite structure (secondary structure and primary flight controls) depend on the extent of damage, the time available to perform the repair, and the time until the next scheduled maintenance visit. In approximately 80 percent of all cases, the damage is covered with adhesive-backed aluminum foil ("speed tape") or temporarily repaired and deferred for a specific time to provide for interim or permanent repair or part replacement. Occasionally, temporary or permanent repairs can be performed by bonding or bolting a sealant-coated metal or precured composite overlay over the damage. Finally, most permanent repairs are accomplished with room-temperature curing, wet lay-up and precured patch techniques. Other permanent repairs use prepreg that cures under vacuum or autoclave pressures at temperatures lower than the cure temperature of the original structure. Repair resins are being developed that have relatively low cure temperatures, but have thermal and environmental resistance similar to higher-temperature curing systems.

The thicker laminate construction used in composite primary structure are not conducive to wet lay-up patch technologies. Thin facesheets on honeycomb panels are currently repaired using bonded scarf patches with a scarf taper of 20:1. For thicker constructions the result would be the removal of a large amount of undamaged material (Bodine et al., 1994). The emphasis in the development of primary structure repairs has therefore been on fastened, precured composite or metallic splice plates, similar to current metal repair techniques. A design for a fairly complex bolted repair is shown in figure 7-2. The issues that must be addressed in these types of repairs include (1) criteria for determining when repairs are required; (2) availability of standardized repair elements; (3) drilled hole quality; (4) ability to restore original strength, durability, and damage tolerance; and (4) ability to match existing contours.

Given typical flight schedules, composite structure repair must be accomplished within eight hours on an overnight layover, otherwise the part would be replaced with a spare. Repairs carried out during an overnight stop (at line stations or hubs) and repairs requiring more-intensive maintenance center rework should follow guidelines established by the manufacturer's structural repair manual and the appropriate industry group, the Commercial Aircraft Composite Repair Committee. Because composite structures are fabricated from a large number of resin/fabric systems from several qualified suppliers worldwide, it is difficult and expensive for airlines to stock a wide variety of repair materials. Accordingly, there is a pressing need for standardization of repair materials and processes.

While next-generation aircraft are expected to use laminated or tailored composite skin structure that is more damage tolerant, there are many lessons that have been learned in maintaining and repairing the thin-skin, nonmetallic honeycomb sandwich constructions on today's aircraft that need to be considered in new aircraft design and materials selection. Maintainable designs need to consider component accessibility, permitted defect levels, and nondestructive testing

TABLE 7-4 Causes of Service Damage to Composite Structure

Damage Mechanism	Affected Components
Mechanical Damage (in flight)	
Hail impact	Radome Engine inlet Upper wing and tail plane fixed panels Flight controls
Bird strike	Engine inlet cowl Radome
Engine disintegration	Engine cowl Fuselage Lower wing and tail plane fixed panels
Tire protector separation	Flaps Lower wing-to-body fairings Landing gear doors
Mechanical Damage (on ground)	
Hail impact	All horizontal surfaces (wing panels, flight controls, upper areas of engine cowl)
Groundhandling equipment	Engine cowl Wing and tail leading and trailing edges Landing gear doors
Mishandling	Engine cowl Access doors
Overload due to actuation system failure	Flight controls Spoilers Thrust reverser
Transport and handling	All removable components
Lightning strike	Radome Leading and trailing edge components (aileron, rudder, elevator, leading-edge fairings) Engine cowl
Overheat	Engine cowl Landing gear door (in case of brake overheat)
Erosion	Radome Engine inlet cowl Leading-edge fairings
Chemical contamination	
Skydrol (hydraulic fluid) leakage	Engine cowl Actuated components (flight controls, spoilers)
Paint stripper	All painted components
Corrosion	All aluminum honeycomb with composite faceskins Improperly isolated aluminum brackets, hinges, etc.

SOURCE: Blohm (1994).

techniques. Nondestructive test indications need to be correlated with structural criteria throughout the life of the aircraft.

Composites must be protected by finishes with resistance to fluid penetration and ultraviolet degradation. Significant costs are incurred by the airlines over the life of an aircraft in the maintenance of protective finishes. The durability of protective finish systems (including aerodynamic surfaces) should be characterized prior to production.

The removal of finishes from composites is a slow and expensive process. Since chemical strippers attack the polymer matrix, airlines generally remove finishes through mechanical abrasion processes. New paint removal processes like laser, heat, frozen carbon dioxide blasting, and wheat starch blasting are being evaluated. Rapid, low-cost, on-aircraft paint removal techniques require implementation if larger areas of composite surfaces are to be accepted on the next-generation aircraft.

Fast, durable, temporary, and permanent field and shop repair procedures should be developed for new composites. This would include development of repair materials, tooling, and processes for high-modulus, high-strength composite skins, metallic and nonmetallic honeycomb sandwich structure, and laminated hybrid and bonded metal structures. These repairs should be accomplished without major disassembly of structure—preferably while working from an exposed exterior or interior surface. It is critical to design for accessibility and interchangeability of parts, especially those for parts that have high damage probability.

SUMMARY

Consideration of aircraft maintenance and repair procedures is a critical part of the development and application of new materials and structures. Previous service experience with metallic and composite structures supports the importance of a maintainable design. The experience of the aging fleet with metallic structures provides lessons in corrosion prevention and control as well as detection and control of multiple-site fatigue damage through appropriate analysis methods, improved component designs, and focused inspection and maintenance. Experience in thin-skin composite components suggests emphasis on robust and durable component design and standardization of repair criteria, materials, and procedures.

Additional experience suggests that the heavy dependence on the costly and intensive inspections to maintain safety in aging airplanes must be reduced further by implementing new and easier to use inspection techniques; utilizing materials with better damage tolerance, toughness, and visible warnings of impending failure; and designing components with adequate access for inspection and maintenance to preclude hidden defects. The reliance on inspection technology can also be reduced by designing structure that is tolerant of

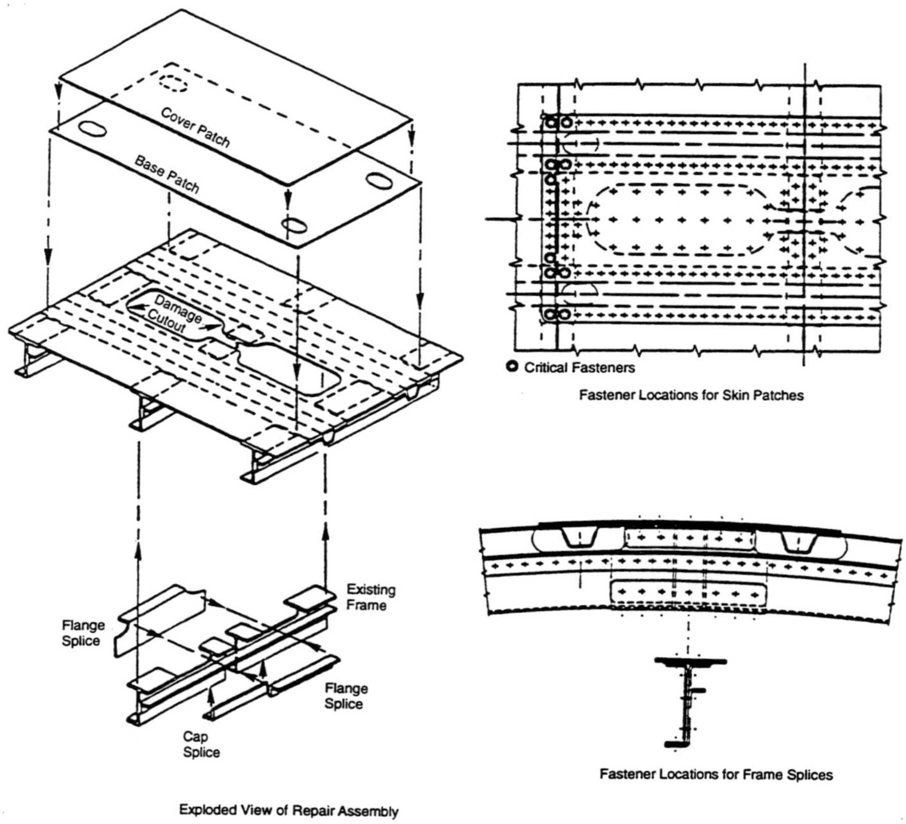

FIGURE 7-2 Bolted splice repair of a composite primary structure panel. Source: Bodine et al. (1994).

undetectable defects or damage; establishing damage limits as well as specific inspection standards and techniques; and developing damage-tolerant, fatigue-resistant structural repairs for new materials along with the means for inspection validation.

The FAA should work with the airline and manufacturing industries to develop common standards and procedures for maintenance and repair of aircraft structure. Application of new materials, processes, and component designs need to be anticipated and accounted for in the FAA's research priorities.

8

Nondestructive Evaluation

Inspections of commercial transports encompass a variety of scheduled services or checks of the aircraft with predetermined work content. These checks occur at defined frequencies, either flight hours or calendar days, depending on the type of aircraft and its typical mission (see table 7-1 for typical schedule of maintenance and service). The airline industry uses many different nondestructive evaluation (NDE) technologies during these service and maintenance operations. The common inspection methods currently used are:

- visual (normal and intensified),
- penetrant,
- audio sonic,
- leak-checking components (hot water immersion),
- X-ray (one or two position),
- eddy current for corrosion,
- ultrasonic (pulse echo or through-transmission),
- resonance testing,
- coating color change for heat damage,
- conductivity check of metal parts for heat damage,
- hardness check in noncritical areas, and
- moisture detector for nonconducting parts.

There are many available references that specifically address application of these common NDE techniques to airframe structures. General overviews of NDE in aerospace have been discussed by Heyman (1989) and Achenbach (1992).

Hagemaier (1988) discussed the organization, procedures, and methodologies to implement NDE into the total product cycle for commercial aircraft. He described the typical process for the selection and development of an inspection process for maintenance activities. Inspection requirements were developed in response to the discovery of cracks or corrosion during routine maintenance. Inspections have also been developed based on structural analyses and on the results from full-scale fatigue and damage tolerance testing. With the introduction of new materials with known durability limitations caused by fatigue or other damage, it is important to determine NDE requirements prior to maintenance or in-service actions.

On-aircraft eddy current inspection is used for the detection of subsurface cracking, which usually occurs at areas of high stress concentration (e.g., splice joints) (Hagemaier et al., 1988). This low-frequency eddy current inspection is performed by scanning the 1.0 kHz reflectance probe over the lower wing skin. The inspector observes changes in the cathode-ray tube display for crack indications. Typically, these traces are difficult to read and interpret. The result of the inspection is classically described as "go" or "no-go" (i.e., either no defect [crack] was detected or there was a defect [crack] detected). There has been little quantitative analysis done in-service to determine the flaw characteristics.

Much of the work concerning aging aircraft has focused on NDE (Bobo, 1989; Kotzian, 1989; Hagemaier, 1990). Three main areas of concern have been identified, including crack detection, corrosion detection, and lack of bond in lap joints. Hagemaier (1990) described NDE techniques that are useful for each of these defects. Unfortunately, all of the techniques described, including eddy current, radiography, and ultrasonics, are time-consuming and sensitive to inspector interpretation.

Inspection issues continue to create problems for the airlines during maintenance of aging aircraft. The most critical issues inhibiting a successful maintenance program include inadequate inspection standards, lack of quantitative defect interpretation, and lack of definitive rejection criteria. In addition, inspections for large-scale parts, such as the wings or fuselage, for the detection of the onset of multiple-site damage, corrosion, or other types of in-service damage is extremely time-consuming and tedious. To detect multiple-site damage and corrosion, commercial transport operators need cost-effective, wide-area inspection methods to detect:

- hidden corrosion,
- small or subsurface cracking (not visually inspectable),
- damage in inaccessible locations, and
- disbonds and weak bonds (Sampath, 1993).

The results of current inspection programs, together with the aging aircraft experience, place strong emphasis on the need for the development of improved standards for NDE methodologies and the specific materials and structural applications. In particular (and with respect to this NRC study) it is important that inspection standards be developed to represent the new materials and structural concepts with the specific defects known to be detrimental to the damage tolerance and durability of the aircraft.

In recent inspection reliability studies conducted by the FAA Technical Center and Sandia National Laboratories, inspection times ranged from 2–5 labor hours per 80 feet of lap slice using a standard eddy current technique. Experience at United Airlines has shown that a typical 747 lap-slice inspection requires four people for two shifts (i.e., 64 labor hours) to complete. Furthermore, inspection times do not include the stripping of paint which is often required. Other inspections, including ultrasonic inspections (low-frequency bond test, pulse echo, and through-transmission contact) require similar times.

The application of instrumental NDE technology only constitutes approximately 10 percent of the inspection accomplished on today's aircraft. However, with the aging of the commercial fleet and the need to detect the onset of multiple-site damage to protect structural safety, the reliance on instrumental NDE methods could increase. The remaining (90 percent) inspections are currently accomplished using visual inspection methods. Visual inspection can be even more tedious and considerably more subjective than other NDE techniques. A further consideration is that extensive disassembly is often required in order to perform visual inspections (Dreher, 1995).

During the past 20 years, research in NDE has been expanding to respond to the needs of all industries, including aerospace. Centers of excellence for NDE research and development have been established at Iowa State University and The Johns Hopkins University. In addition, many other well-known universities around the world have established NDE as a significant research area in their schools of engineering or applied science.

A comprehensive overview of NDE research specifically for aerospace is presented by Achenbach and colleagues (ASME, 1992). The following sections provide a brief overview of technologies that are projected by the committee to have the greatest impact on next-generation commercial aircraft. It is assumed that current NDE methods such as visual, ultrasonics, eddy current, and radiography will continue to be used.

THERMAL METHODS

Thermal inspection comprises all methods in which heat-sensing devices are used to measure temperature variations in the system being inspected. An overview of thermal inspection is presented by Hardy and Bolen (1989). A more advanced overview by Henneke and Tang (1992) specifically described thermal-wave imaging, both real-time video and long-time scanning.

Infrared thermography provides a rapid means for inspecting large surfaces (Henneke and Tang, 1992). The surface of a large component can be scanned quickly to identify problem areas which can be subsequently inspected in more detail to determine quantitative results. Recent studies have identified infrared thermography to be useful to detect defects and damage in composite materials. In addition, thermographic testing was determined to be a good method for detecting defects in foam-core structures (Vikstrom, 1989).

The Lawrence Livermore National Laboratories investigated and demonstrated the application of dual-band infrared imaging as a dynamic thermal tomography tool for wide-area inspection of the Boeing 737 (Del Grande et al., 1993). However, this technique is still an emerging technology and requires additional work to verify that the implications based on laboratory calibration standards are consistent with the effects of corrosion on actual aircraft structures.

ACOUSTIC EMISSION

Acoustic emissions are transient elastic waves emitted when there is a sudden change of stress in materials. These waves radiate through the structure and can be detected on the surface using one or multiple sensors. Acoustic emission (AE) has been investigated for many years with its application as a useful NDE method beginning in the late 1940s and early 1950s (Drouillard, 1988). An extensive overview of AE measurements as well as recommendations for future work has been prepared by Sachse and Gorman (1992). The characteristics of the AE response of a wide variety of aerospace alloys, polymer-matrix and metal-matrix composites, as well as the correlation of AE data with load, strain, and deflection nonlinearities in composite airframe structural components, have been reviewed and summarized by McBride et al. (1987).

The challenges and approaches to monitoring AE signals in aircraft structures have been described by McBride and Maclachlan (1984). AE data includes noise from airframe rubbing and fretting and rapid changes in structural loading during flight. The in-flight monitoring of AE signals, however, seems possible because noise far from the sensor results in mostly low-frequency signals, and engine-generated noise is not a problem above 300 kHz. Unfortunately, the number of signals generated by slow crack growth is several orders of magnitude lower than the number of airframe noise signals (McBride and Maclachlan, 1984).

Another approach has been to exploit pattern recognition techniques to identify crack growth signals and separate them from fretting noise (Friesel, 1989). Waveform classification features that allowed discrimination between AEs from crack growth and from fretting in a thick-jointed aluminum plate have been developed. This research was performed in the laboratory using a test specimen (7075-T651 aluminum) with a pin-loaded joint to enable the generation of fretting noise.

The technique of AE must be more fully understood, including the emission source, structure, sensors, and analyzing system to make it a more practical and reliable inspection method for materials and structures (Sachse and Gorman, 1992). However, AE has many advantages that make it a candidate for a fast, "whole-field" technique, including the capability to detect cracks that are actively growing. The capability to distinguish the defect noise from other airframe noise continues to be a major issue that will require additional research.

LASER NDE METHODS

There are several types of laser NDE methods: acoustic holography, shearography, laser ultrasonics, optical holography, and acoustic microscopy. Some of these techniques have the advantage of not requiring contact, allowing a "full-field" view of the structure.

Shearography, an interferometric method, is receiving extensive attention as a potential technique for airframe structures. A major limitation is the need to apply load to the structure during the inspection process. Hung (1989) has presented a detailed discussion of shearography.

Acoustic holography and optical holography (ASM, 1989) use the same principles because the laws of interference and diffraction apply to all forms of radiation. Differences between the methods are due to the method of recording the output. Optical holography, a noncontact method, has been used to detect disbonds in honeycomb sandwich structures. Acoustic holography requires a liquid interface to transfer the acoustic waves into the material being inspected. Therefore, acoustic holographic methods are used on smaller samples either in production or in the laboratory.

Recent research by Ferraro and colleagues (1994) has investigated the use of holography interferometry for large composite aircraft parts. This study evaluated the physical mechanisms by which the flaws are detected. Particular attention was focused on the inspection of sandwich composite parts made of two skins of carbon-fiber-reinforced epoxy bonded to aramid (Nomex®) honeycomb core. This research resulted in an optimized technique based on the physical mechanisms of disbond detection. However, like most new NDE methods, development of these laser-based methods has been conducted primarily in research environments. The transfer of the technology into actual field conditions has been limited, and little effort has been extended for full implementation (e.g., training and certification of inspectors and maintenance of equipment).

OTHER NDE METHODS

There are many other NDE methods being studied, including advances in ultrasonics, electromagnetic, radiographic, optical enhancers, microwaves, and others. Other improvements are being researched in signal processing, analysis, and enhancement. The availability of computers that are compact, fast, and portable is increasing the use of digital technology for real-time inspections. However, the same NDE technologies used 10–20 years ago are still being used today for the majority of inspections. Recent advances in NDE technology have not been successfully transferred from the laboratory into field service.

As new materials and structures are introduced into next-generation aircraft, it is imperative that the NDE techniques required to support them are transferred successfully into the maintenance and repair facilities.

Aging aircraft experience has shown that NDE techniques require standards to improve the overall repeatability and reproducibility of the inspection. Inspections are now costly and time-consuming. Cost-effective, quantitative NDE methodologies to enhance maintenance inspections should be developed and demonstrated for next-generation aircraft.

NDE AS AN ENGINEERING TOOL

NDE technology has been principally practiced as an empirical technology, not as a quantitative engineering discipline. In general, NDE methods have been specified for material and component inspection requirements to maintain the necessary quality. In most industries—including the aircraft industry—the inspection requirements are defined in a specification that describes the sensitivity level of the inspection method as well as the rejectable flaw size.

Damage-tolerant assessments, however, rely heavily on knowledge regarding the flaw size that can be detected. In the 1970s, the U.S. Air Force introduced the damage-tolerant assessment concept for the military aircraft industry. To determine flaw-size detectability for in-service inspection, Lockheed conducted the first evaluation of NDE methods (Lewis et al., 1979). The evaluation concluded that the overall reliability of NDE performed by the Air Force fell significantly below the anticipated and needed capabilities. NDE reliability analysis or probability of detection has been thoroughly discussed in the literature and is commonly used in today's industry to determine inspection capability (Berens, 1989). The difficulty in using the present NDE techniques is that the cost of conducting a reliability study is extremely prohibitive for the commercial airline industry. Furthermore, the creation of representative samples with known defects is extremely difficult, as well as expensive. For example, U.S. Air Force studies in the 1980s for gas turbine engines cost over $1 million per study (Cooper, 1995), and recent efforts by the Nuclear Regulatory Commission to develop a set of parts with "real" defects cost in excess of $10 million (Turnbow, 1995).

Significant advances have been made recently that involve the development of additional engineering tools to "predict" the response of the NDE measurement process (Thompson, 1992). Quantitative models of the NDE measurement processes (i.e., a "measurement" model for each of the various inspection technologies) will provide an engineering basis for subsequent NDE measurement.

The purpose of the measurement model is to predict the NDE system's response to specific defects in a given material or structure. The model must include (1) the geometric details of the part; (2) the material(s) of construction; (3) the inspection system details (i.e., sensor, instrument, orientation, etc.), and (4) the NDE energy model including the generation, propagation, and reception of the energy (i.e., ultrasound waves, eddy currents, X-rays, etc.). The measurement model must also consider characteristics of the energy interaction with the material, noise sources, energy losses due to material anisotropy, and any other variations caused by the entire system.

Several NDE measurement models have been developed over the past several years, including models for ultrasonics, eddy current, and radiography (Thompson, 1992). One useful application of these measurement models is the prediction or simulation of probability of detection (Gray et al., 1989). Figure 8-1 shows the results of a detectability simulation of circular cracks at three different depths and for two scan plans. In the case shown, the geometry of the part focuses the acoustic beam at the intermediate depth (25 mm [1 in.]), resulting in a reduced beam width and lower probability of detection.

The development of NDE measurement models for the commercial aircraft industry will significantly improve the current NDE state-of-the-art practice. The models, if developed, will provide:

- prediction of NDE reliability for both production and in-service inspection without costly and time-consuming experimental demonstration programs;
- improvements of inspection procedures; and
- integration of the NDE reliability with the damage-tolerant and durability assessment to optimize total life for a given system.

SUMMARY

The emphasis for future NDE developments should stress (1) reliability of defect detection, (2) cost-effectiveness, and (3) ease of implementation in field environments. The FAA can play a pivotal role in assessing the capabilities of new methods to detect defects critical in aircraft operation, developing inspection standards and reliability characterization, and transferring the technology into fleet applications. Improvements in NDE standards and methods are critical in the monitoring of the commercial aircraft fleet and must keep pace with new developments in materials and structures technology.

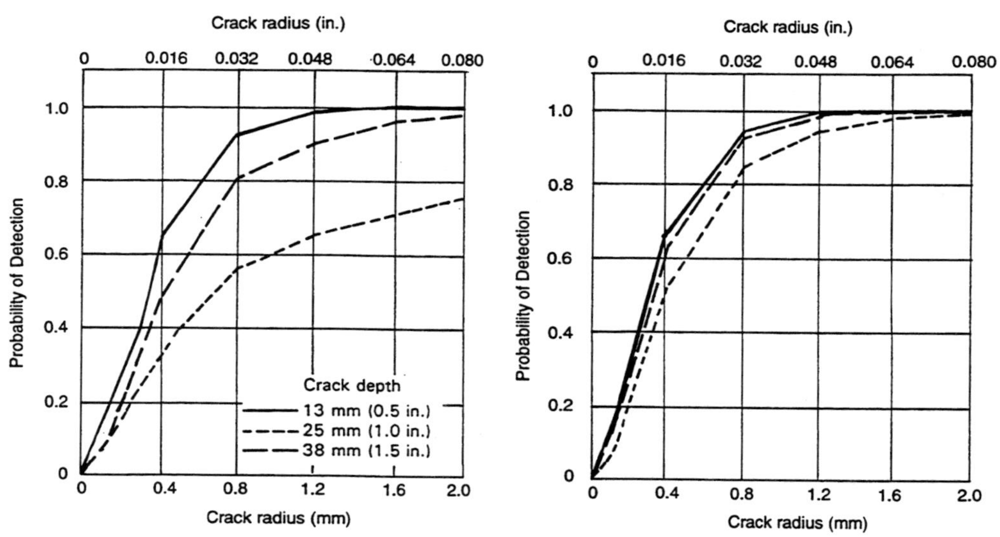

FIGURE 8-1 Probability of detection simulations for ultrasonic detection of circular cracks at different depths below a component surface for two scanning plans. Source: Gray and Thompson (1986). With permission of Plenum Publishing Corp.

V

Conclusions and Recommendations

9

Committee Findings

The major objective of this study was to identify engineering issues related to the introduction of new materials and their expected effect on the life-cycle durability of future civil transport aircraft. The committee investigated the likely new materials and structural concepts for next-generation commercial aircraft and the key factors influencing application decisions. Based on these predictions, the committee identified and analyzed the design, characterization, monitoring, and maintenance issues that appear to be most critical for the introduction of advanced materials and structural concepts.

The committee's findings are organized into three sections:

- General conclusions, identifying the influencing factors and new materials, processes, and structural concepts likely to see application on next-generation commercial transport aircraft.
- A description of the roles of the Federal Aviation Administration (FAA) and of other government, industry, and academic organizations in technology development for next-generation aircraft.
- Identification of research opportunities and the committee's recommendations in three primary areas: (1) materials, manufacturing, and structural concepts; (2) methods for assessment of structural performance; and (3) inspection, maintenance and repair.

CONCLUSIONS

There has been significant recent progress in the introduction of new materials and structural designs in commercial transport aircraft. For example, toughened, polymeric composite primary structure was introduced on the Boeing 777 empennage, and structural aluminum castings were introduced on the Airbus A340. Aircraft designers continue to apply new materials and structural concepts to provide benefits in performance, durability, compliance with environmental regulations, and most recently, acquisition and maintenance costs. The use of new materials and structures will continue to expand on next-generation aircraft.

As noted, the current turbulent, economic climate affecting the airline, manufacturer, and materials industries has significantly changed the application criteria for advanced materials. As a result, materials performance is no longer the only primary driver for materials selection. Aircraft manufacturers are responding to airline concerns about reducing overall costs, including the costs of acquisition and maintenance. The result is incremental, evolutionary material changes rather than revolutionary ones. The principal barriers to increased use of new high-performance materials are:

- Costs (acquisition, manufacturing, certification, life cycle) relative to benefits, vis-à-vis "old" designs based on "old" materials.
- Incomplete understanding of basic failure mechanisms and their interactions in advanced materials—particularly composite materials—and their structures.
- Industrial conservatism engendered by perceptions of technological risk. The industry lacks the experience that would allow the understanding of the durability of advanced materials and structures.
- The state of the materials supplier base. The specialty materials industries find it difficult to make the long-term financial commitment needed to undertake a major development program.

Principal "new" airframe materials expected to realize increased use in the next generation of advanced civil aircraft include polymer-matrix composite primary structure (laminates, tailored forms, woven and sewn three-dimensional configurations, automated tape and tow placement) and advanced metals and alloys (tough aluminum, high-yield aluminum, aluminum-lithium, high-strength titanium, and high-strength steel). Continued incremental improvements in metal alloys represents a low technological risk in that the design tools and characterization methods, analytical tools, and design issues differ little from current procedures. On the other hand, the expanded application of composites in primary structure and the application of innovative composite and metals processes (e.g., net-shape processing) requires substantial improvements in characterization and analysis methods and thus represents a higher technological risk and is unlikely to proceed as rapidly.

In contrast, given the emphasis on incremental technology advances and total costs, the committee does not foresee significant application of metal-matrix composites in the airframes of next-generation transports. New advanced materials applications in major subsystems such as carbon/carbon composites in brakes and ceramic nozzles in auxiliary power units have been more aggressive. These components have not been included in this study and may be worthy of future consideration by the FAA.

Increasingly, airframe manufacturers are using an integrated product development approach that considers such factors as producibility, cost, nondestructive evaluation (NDE) methods and criteria, and repair and maintenance issues and involves airline designers, manufacturers, and suppliers from the outset of development programs.

Commercial aircraft are built and operated on a global basis with international teaming of manufacturers, suppliers, and fabricators. Accordingly, the development and harmonization of international standards for materials and processes, testing and evaluation, NDE, and repair and maintenance procedures are critical to developing and commercializing new materials and structures technology. In spite of the international teaming involved in developing a new aircraft, the manufacture and operation of commercial transports remains an extremely competitive business. The committee believes that competitive pressures will continue to influence the criteria for the application of new materials and processing technology.

ORGANIZATIONAL ROLES

There are a number of organizations—airlines, aircraft manufacturers, suppliers, the FAA, university researchers, the National Aeronautics and Space Administration (NASA), and the U.S. Department of Defense—that are involved in the development and application of new materials and structures technology for aircraft. The airlines are responsible for establishing performance needs that will keep them competitive and for establishing and implementing inspection and maintenance procedures required to operate the aircraft in a safe and cost-effective way. The aircraft industry and their suppliers are ultimately responsible for the development, evaluation, application, and validation of new technologies and operating procedures for production-scale utilization. Industry focus is on short-term developments and technologies that will allow them to remain competitive.

While the industry performs a significant amount of research and development, it does not generally perform basic, precompetitive research. NASA has been described as "the only organization in the United States with both the capability and the mandate to perform the basic research as well as the ground and flight testing necessary to validate new concepts to the extent that they can begin to be incorporated into commercial aircraft" (NRC, 1992). NASA has a history of success in this type of activity, including composite long-term testing and flight-service evaluations and the current Advanced Composite Technology program discussed in chapters 2 and 4.

University research organizations work in concert with industry and government efforts to provide basic research and development tools, evaluation and analysis, and workforce education and training.

The continuing research and development activity in materials and structures conducted by the FAA is fueled by the responsibilities and mandates of the organization. First, the FAA must continue to keep abreast of technology needed to support their Aircraft Certification Service and Flight Standards Service in the certification of new aircraft and the effort to monitor the safety of the aircraft fleet. Second, the FAA has a mandate to undertake research to develop technologies to assess the effect of aircraft design, maintenance, testing, wear, and fatigue and to develop improved technology and practices for maintenance (including NDE) (P.L. 100-591) and to develop technology to assess the risk of and prevent failures or malfunctions that would lead to catastrophic failure (P.L. 101-508). In addition, the FAA is the logical organization to work with industry in the harmonization of standards and operations and maintenance procedures with foreign industries and agencies.

RECOMMENDATIONS

As described in the previous section, technological advances are brought about through the concerted efforts of airline, industry, academic, and government organizations. In forming their recommendations, the committee identified the technologies that are likely to be involved in the development of next-generation aircraft and outlined the work required to bring about those developments. The recommendations are directed toward all of the organizations involved in new materials applications, but also specifically toward what the FAA role should be in these developments.

- In general, the committee recommends that the FAA remain involved in all stages of the technology development process, with emphasis on work related to aircraft safety, operations, maintenance, and nondestructive evaluation.

Materials, Manufacturing, and Structural Concepts

The future improvements in aircraft structural components will continue to be based on factors related to materials selection, analytical methods, structural concepts, and processing innovations. With the increased emphasis on affordability, it is probable that fewer new materials will be developed. On the other hand, robust and cost-effective processing methods as well as compliance with environmental regulations will become paramount issues to provide lower costs. Forming, joining, and finishing processes that contribute to reduced labor hours and fewer detailed part counts will have a major impact on reduced overall acquisition costs.

Specifically, the committee envisions continued improvements in the performance and durability of metal alloys and polymeric composites. Materials and structures more conducive to low-cost processes such as casting, high-speed machining, and superplastic forming/diffusion bonding of metals and fiber placement, resin transfer molding, and nonautoclave processing of composites will be emphasized in the future.

The committee recommends that the FAA work with industry, government, and academic organizations in the development of new materials, processing, and structure technology by the following guidelines:

- Keep abreast of innovative materials processing technologies that provide methods for low-cost fabrication of aircraft structure. Emphasis should be placed on the understanding of new product forms, processing methods, and thermal treatments and their possible effects on materials performance.
- Support the development of emerging process modeling techniques for definition of processing parameters and requirements.
- Establish and maintain databases of material and structural properties resulting from the candidate processing methods. The databases should include test methods, physical and mechanical properties, failure modes, and influences of probable defects and manufacturing processes on property behavior.
- Work with the materials, manufacturing, and airline industries to develop industrywide standards to improve consistency in the final products, especially with the increasing globalization of materials availability.
- Participate in industry- and NASA-sponsored flight hardware demonstration programs for the introduction of new materials, manufacturing processes, and structural concepts in high-risk applications. FAA emphasis should be on validation of inspection and repair techniques and in the development of technology needed to certify and monitor these structures.

Methods for Assessment of Structural Performance

Current structural design and analytical procedures used by the aircraft industry are largely semiempirical, even though significant improvements have occurred in structural analysis methodology over the last two decades. Accurate, finite element analysis methods are used routinely for predicting the stress, strain, and displacement fields in complex structural geometries. However, the reliable prediction of structural failure modes, ultimate strength, residual strength, and fatigue life has remained elusive to the structural engineer. The current standard practice relies heavily on extensive testing at the coupon, subelement, element, subcomponent, component, and full-scale levels. Design details are frequently optimized through test programs. Scale-up effects are handled through a building-block approach that relies on testing to verify the anticipated structural performance at each scale level. While the committee anticipates that this building-block approach to structural design will continue indefinitely, a more rigorous, analytical prediction methodology will greatly improve the process of introducing new materials into airframe primary structure.

The committee recommends that the FAA work with other industry, government, academic organizations in the development of improved analytical methods by the following guidelines:

- Support development and facilitate implementation of advanced analytic and computational methodology to predict residual strength as a function of time.
- Support programs to improve the understanding of basic failure mechanisms in advanced materials and their structures. Include the interactions of the various failure modes manifested at the various length scales—from material to structural levels.

Inspection, Maintenance, and Repair

The successful application of new materials and structural concepts relies on an effective maintenance program that is cost-effective, while ensuring passenger safety. The aging aircraft experience has provided the airline industry with significant lessons learned for inspection and repair technologies. These lessons provide a framework for improving inspection and repair processes for next-generation materials. Major issues that continue to limit the effectiveness of an aircraft maintenance program are poor structural inspection standards, inadequate defect indication interpretation, unreliable inspection techniques, high cost of new NDE methods, and limited linkage with design analyses and NDE results. The leadership of the FAA and the continued participation of airlines and manufacturers in developing and implementing improved maintenance and inspection methods is crucial.

The committee recommends that the FAA take a leadership position in the development of improved inspection and maintenance methods by the following guidelines:

- Support the development of improved standards for NDE methodologies and their specific materials and structural applications, especially through participation in industry- and NASA-sponsored component development and flight hardware demonstration programs for the introduction of new materials, manufacturing processes, and structural concepts.

- Support the development of cost-effective, quantitative NDE methodologies for in-service inspection of airframe materials and structures. Emphasize improved defect detection reliability, cost-effectiveness, and ease of implementation in field environments. Particular attention should be given to rapid, wide-area inspection with limited or one-sided access.
- Develop improved analytic methods to determine NDE reliability and inspectability of materials and structures to support damage tolerance and durability analyses.
- Support the development of real-time repair and maintenance processes for materials and structures that use the results from quantitative NDE methods and computational analyses.

References

Achenbach, J.D. 1992. Introduction: the science and technology of NDE. Pp. 153–166 in Flight-Vehicle Materials, Structures, and Dynamics–Assessment and Future Directions. Volume 4: Tribological Materials and NDE, R.L. Fusaro and J.D. Achenbach, eds. New York: American Society of Mechanical Engineers.

AFML (U.S. Air Force Materials Laboratory). 1973. Summary of Air Force/Industry Manufacturing Cost Reduction Study. Technical Memorandum AFML-TM- LT-73-1, January. Wright Patterson Air Force Base, Ohio: AFML.

AGARD (Advisory Group for Aerospace Research and Development). 1991. Handbook on Advanced Casting. AGARD-AG-299. Neuilly Sur Seine, France: North Atlantic Treaty Organization.

ASM (ASM International). 1989. Metals Handbook. Volume 17: Nondestructive Evaluation and Quality Control. Metals Park, Ohio: ASM International.

ASME (American Society of Mechanical Engineers). 1992. Flight-Vehicle Materials, Structures, and Dynamics: Assessment and Future Directions. Volume 4: Tribological Materials and NDE, R.l. Fusaro and J.D. Achenbach, eds. New York: ASME.

ASTM (American Society for Testing and Materials). 1994. Standard Test Method for Mode I Interlaminar Fracture Toughness of Unidirectional Fiber-Reinforced Polymer Matrix Composites. ASTM D-5528. Philadelphia, Pa.: ASTM.

ATA (Air Transport Association of America). 1993. State of the U.S. Airline Industry: A Report on Recent Trends for U.S. Air Carriers. Washington, D.C.: ATA.

ATA (Air Transport Association of America). 1995. Air Transport 1995: The Annual Report of the U.S. Scheduled Airline Industry. Washington, D.C.: ATA.

AWST (Aviation Week and Space Technology). 1994a. Boeing refines designs for 600-seat NLA. Aviation Week and Space Technology 141(21):48–53.

AWST (Aviation Week and Space Technology). 1994b. Airbus weighs four A3XX versions. Aviation Week and Space Technology 141(21):54.

AWST (Aviation Week and Space Technology). 1994c. Douglas awaits chance to revive jumbo plans. Aviation Week and Space Technology 141(21):57.

AWST (Aviation Week and Space Technology). 1995. Zero accidents, zero excuses (editorial). Aviation Week and Space Technology 142(4):70.

Bania, P.J. 1989. Ti-1100: a new high temperature titanium alloy. Pp. 825–830 in the Sixth World Conference on Titanium, P. Lacombe, R. Tricot, and G. Beranger, eds. Cedex, France: Société Française de Metallurgie et des Materiaux.

Bania, P.J. 1993. Beta titanium alloys and their role in the titanium industry. Pp. 3–24 in Beta Titanium Alloys in the 1990s, D. Eylon, R.R. Boyer, and D.A. Koss, eds. Warrendale, Pa.: TMS.

Bania, P.J. 1994. Beta titanium alloys and their role in the titanium industry. Pp. 720 in Beta Titanium Alloys, A. Vassel, D. Eylon, and Y. Combres, eds. Paris, France: Société Française de Metallurgie et des Materiaux.

Bania, P.J., A.J. Hutt, R.E. Adams, and W.M. Parris. 1993. A new low cost titanium alloy. Pp. 2787–2794 in Titanium '92 Science and Technology, F.H. Froes and I.L. Caplan, eds. Warrendale, Pa: TMS.

Berens, A.P. 1989. NDE reliability data analysis. Pp. 689–701 in Metals Handbook. Volume 17: Nondestructive Evaluation and Quality Control. Metals Park, Ohio: ASM International.

Blankenship, C.P., Jr., and E.A. Starke, Jr. 1992. Fracture behavior of aluminum-lithium-X alloys. Pp. 187–201 in Aluminum-Lithium, Vol. 1. M. Peters and P.J. Winkler, eds. Verlag, Germany: Informationsgesellschaft.

Blanton, R.C., R.W. Englebart, C.R. Green, J.K. Lehman, and G.G. Wombacher. 1995. GLCC RTM Program, Tool Design Manual. Great Lakes Composites Consortium (GLCC), Navy RTM Manufacturing Technology (MANTECH) Program, TDL-91-07, Cooperative Agreement N00014-90-CA-0002. Washington, D.C.: Office of Naval Research.

Blohm, C. 1994. Advanced Composites Structures: In-Service Experience in the View of a Carrier. Paper presented at the Workshop on Long-Term Aging of Materials and Structures, National Materials Advisory Board, National Research Council, Washington, D.C., August 10–12.

Bluck, R.M. 1969. High speed bias weaving and braiding. U.S. Patent 3,426,804.

Bobo, S. 1989. Nondestructive inspection of aging aircraft. Pp. 18–33 in FAA Aging Aircraft Workshop Proceedings. Columbus, Ohio: American Society for Nondestructive Testing.

Bodine, J.B., B.W. Flynn, M.H. Gessel, K.H. Griess, L.B. Ilcewicz, G.D. Swanson, M.F. Nahan, and J.R. Epperson. 1994. Repair of Composite Fuselage Panels. Paper presented at the Fifth NASA/DOD Advanced Composite Technology Conference, Seattle, Wash., August 22–25.

Boeing. 1994a. Ramp rash. Boeing Airliner, April–June.

Boeing. 1994b. Cargo Facility and Equipment Handling. Document D6-26001. Seattle, Wash.: Boeing.

Boyer, R.R. 1980. Design properties of a high strength titanium alloy, Ti-10V-2Fe-3Al. Journal of Materials 32:61–65.

Boyer, R.R. 1993. Applications of beta titanium alloys in airframes. Pp. 335–346 in Beta Titanium Alloys in the 1990s, D. Eylon, R.R. Boyer, and D.A. Koss, eds. Warrendale, Pa.: TMS.

Boyer, R.R. 1994. Aerospace applications of beta titanium. Pp. 253–264 in Titanium Alloys, A. Vassel, D. Eylon, and Y. Combres, eds. Paris: Société Française de Metallurgie et des Materiaux.

Boyer, R.R., R. Bajoraitis, D.W. Greenwood, and E.E. Mild. 1984. Ti-3Al-8V-6Cr-4Mo-4Zr wire for spring application. Pp. 295–305 in Beta Titanium Alloys in the 1980s, R.R. Boyer and H.W. Rosenberg, eds. Warrendale, Pa.: TMS.

Brookstein, D.S. 1991. Comparison of multilayer interlock braided composites with other 3D braided composites. Pp. 141–150 in Proceedings of the 36th International SAMPE Symposium. Covina, Calif.: Society for the Advancement of Materials and Process Engineering.

Bucci, R.J., L.N. Mueller, L.B. Vogesang, and J.W. Gunnick. 1989. ARALL laminates. Treatise on Materials Science and Technology 31:295–322.

Carey, R.S, R.R. Boyer, and H.W. Rosenberg. 1985. Fatigue properties of Ti-10V-2Fe-3Al. Pp. 1261–1267 in Titanium Science and Technology, G. Lutjering, U. Zwicker, and W. Bund, eds. Deutsche Gesellschaft fur Metallkunde. Oberursel, Germany: eingetragener Verein.

Cawley, P. 1992. NDT of adhesive bonds. Pp. 349–363 in Flight-Vehicle Materials, Structures, and Dynamics – Assessment and Future Directions. Volume 4: Tribological Materials and NDE, R.L. Fusaro and J.D. Achenbach, eds. New York: American Society of Mechanical Engineers.

Chu, S.G., H. Jabloner, and B.J. Swetlin. 1987. Epoxy thermosets having improved toughness. U.S Patent 4,656,208.

Clark, L., C. Harberg, and R. Boyer. 1995. F-22 materials applications. Paper presented at Aeromat 95, Anaheim, Calif., May 9.

Cogswell, F.N. 1991. Introduction. Pp. 1–25 in Thermoplastic Composite Materials, L.A. Carlsson, ed. Amsterdam, The Netherlands: Elsevier.

Cogswell, F.N. 1992. Thermoplastic Aromatic Polymer Composites. Stoneham, Mass.: Butterworth-Heinemann.

Collings, E.W. 1984. The Physical Metallurgy of Titanium Alloys. Metals Park, Ohio: ASM International.

Cooper, T. 1995. Personal communication, V. Panhuise with T. Cooper, Air Force Wright Laboratories.

Crowe, R.C. 1994. Smart Materials and Structures. Presentation to the Committee on New Materials for Advanced Civil Aircraft, National Materials Advisory Board, National Research Council, Washington, D.C., December 2.

Curtis, D., and J. Lewis. 1992. The FAA aging airplane program plan for transport aircraft. Pp. 321–325 in the Proceedings of the International Conference on Aging Aircraft and Structural Airworthiness, NASA 3160, Washington, D.C., Nov. 19–21, 1991. Washington, D.C.: National Aeronautics and Space Administration.

Davies, D.P. 1993. Effect of heat treatment on the mechanical properties of Ti-10V-2Fe-3Al for dynamically critical helicopter components. Pp. 1851–1858 in Titanium '92 Science and Technology, F.H. Froes and I.L. Caplan, eds. Warrendale, Pa.: TMS.

Del Grande, N.K., D.W. Dolan, P.F. Durbin, M.R. Gorvad, and A.B. Shapiro. 1993. Dynamic thermal tomography for nondestructive inspection of aging aircraft. Pp. 66–77 in the Proceedings of SPIE (The International Society for Optical Engineering) Conference on Nondestructive Inspection of Aging Aircraft, V. 2001, San Diego, Calif.

DeRosa, R. 1995. Personal communication, M. Kuperman with R. DeRosa, United Airlines, January.

DeVault, J.B. 1993. The Future of the Advanced Composite Industry. Paper presented at the Annual Research Symposium of the Center for Composite Materials, University of Delaware, Newark, September 29–30.

Dexter, H.B., and D.J. Baker. 1994. Flight service environmental effects on composite materials and structures. Advanced Performance Materials 1(1):51–85.

Dost, E.F., W.B. Avery, S.R. Finn, L.B. Ilcewicz, D.B. Scholz, and R.E. Wishart. 1993. Impact damage resistance of composite fuselage structure, part 3. Paper presented at the Fourth ACT (Advanced Composite Technology) Conference, Salt Lake City, Utah, June 7–11. Seattle, Wash.: Boeing Commercial Airplane Group.

Dreher, T. 1995. Personal communication, M. Kuperman with T. Dreher, United Airlines, March.

Drouillard, T.F. 1988. Introduction to acoustic emission. Materials Evaluation 46(2):174–180.

Duerig, T.W., and J.C. Williams. 1984. Overview: microstructure and properties of beta titanium alloys. Pp. 19–67 in Beta Titanium Alloys in the 1980s, R.R. Boyer and H.W. Rosenberg, eds. Warrendale, Pa.: TMS.

Edwards, T. 1994. Personal correspondence with Tom Edwards, United Airlines.

Eylon, D., W.J. Barice, R.R. Boyer, L.S. Steele, and F.H. Froes. 1988. Casting of high strength beta titanium alloys.

Pp. 655–660 in the Proceedings of the Sixth World Conference on Titanium, P. Lacombe, R. Tricot, and G. Beranger, eds. Cedex, France: Société Française de Metallurgie et des Materiaux.

FAA (Federal Aviation Administration). 1991. Aircraft Safety Research Plan. Atlantic City, N.J.: FAA Technical Center.

Fanning, J.C. 1993. Ti metal 21S property data. Pp. 397–410 in Beta Titanium Alloys in the 1990s, D. Eylon, R.R. Boyer, and D.A. Koss, eds. Warrendale, Pa.: TMS.

Fanucci, J.P., S.C. Nolet, C. Koppernaes, and Y.R. Kim. 1989. Measurement of internal die pressure distributions during pultrusion of thermoplastic composites. Pp. 50–64 in International SAMPE Technical Conference, Vol. 22. Covina, Calif.: Society for the Advancement of Materials and Process Engineering.

Fanucci, J.P., Y.R. Kim, C. Koppernaes, and S.C. Nolet. 1991. Product and process development of 1,000 foot underwater towed cable system. Pp. 243–268 in Composites in Manufacturing: Case Study, A.B. Strong, ed. Dearborn, Mich.: Society of Manufacturing Engineers.

Ferraro, P., A. Ferraiuolo, S.I. Imaparato, C. Voto, and K.A. Stetson. 1994. On the holographic detection of core-to-skin disbonds in composite sandwich structures. Materials Evaluation 52(12):1376–1381.

Florentine, R.A. 1982. Apparatus for weaving a three-dimensional article. U.S. Patent 4,312,261.

Flynn, B.W., M.R. Morris, S.L. Metschan, G.D. Swanson, P.J. Smith, K.H. Griess, M.R. Schramm, K.S. Willden, and R.J. Humphrey. 1993. Global Cost and Weight Evaluation of Fuselage Keel Concepts. NASA Contractor Report 4541, December. Hampton, Va.: NASA Langley Research Center.

Fowler, G., and S. Phifer. 1993. Resin transfer molding for high fiber/low void content. Pp. 471–476 in 38th International SAMPE Symposium, Vol. 38. Covina, Calif.: Society for the Advancement of Materials and Process Engineering.

Friesel, M.A. 1989. Application of signal analysis to acoustic emission from a cyclically loaded aluminum joint specimen. Materials Evaluation 47(7):842–848.

Gao, Z., and K.L. Reifsnider. 1993. Micromechanics of tensile strength in composite systems. Pp. 453–470 in the 4th Symposium on Composite Materials, ASTM STP 1156. Philadelphia, Pa.: American Society for Testing and Materials.

Gerber, F. 1990. Applications of SPF/DB to the F-22. Presented at the Clinic on Superplastic Forming and Diffusion Bonding. Society of Mechanical Engineers. June 14–15.

Goldsworthy, B., and J. Martin. 1985. Pultruded Composites—A Blueprint for Market Penetration: Part A. Presentation at the 40th Annual Conference of the Reinforced Plastics and Composites Institute, Society of the Plastics Industry, January 28–February 1.

Goodman, D.L., D.L. Birx, and G.R. Palmese. 1996. Composite Energy Curing with High-Energy Electron Beams. Paper to be presented at the 41st International SAMPE Symposium and Exhibition, Anaheim, Calif., March 25–28.

Gray, J.N., T.A. Gray, N. Nakagawa, and R.R. Thompson. 1989. Models for predicting NDE reliability. Pp. 702–715 in Metals Handbook. Volume 17: Nondestructive Evaluation and Quality Control. Metals Park, Ohio: ASM International.

Gray, T.A., and R.B. Thompson, 1986. Use of models to predict ultrasonic NDE reliability. Pp. 911–918 in Review of Progress in Quantitative Nondestructive Evaluation, Vol. 5, D.O. Thompson and D.E. Chimenti, eds. New York: Plenum Press.

Gregory, M.A., and G. Roebrocks. 1991. Fiber Metal Laminate: A Solution to Weight, Strength, and Fatigue Problems. Paper presented at the 30th Annual Conference of Metallurgists. Ottawa, Canada: Metallurgical Society of CIM.

Hagemaier, D.J. 1990. Nondestructive inspection of aging aircraft structure. Materials Evaluation 48(12):1499–1512.

Hagemaier, D.J., B. Bates, and A. Steinberg. 1988. On-aircraft eddy current subsurface crack inspection. Materials Evaluation 46(4):518–522.

Hardy, G., and J. Bolen. 1989. Thermal inspection. Pp. 396–404 in Metals Handbook, Vol. 17. Metals Park, Ohio: ASM International.

Hart-Smith, L.J. 1990. Rating and comparing structural adhesives: a new method. Pp. 471–476 in Engineered Materials Handbook. Volume 3: Adhesives and Sealants, H.F. Brinson, Technical Chairman. Materials Park, Ohio: ASM International.

Henneke, E., and B.S. Tang. 1992. Thermal methods of NDE and quality control. Pp. 223–245 in Flight-Vehicle Materials, Structures, and Dynamics: Assessment and Future Directions. Volume 4: Tribological Materials and NDE, R.L. Fusaro and J.D. Achenbach, eds. New York: American Society of Mechanical Engineers.

Heyman, J.S. 1989. NDE in aerospace—requirements for science, sensors, and sense. IEEE Transactions on Ultrasonics, Ferroelectrics, and Frequency Control 36:581–586.

Hirschbuehler, K.R. 1985. An improved 270°F performance interleaf system having extremely high impact resistance. SAMPE Quarterly 17:46.

Holmes, S.T., S.M. Andersen, J.W. Gillespie, Jr., and S.H. Munson-McGee. 1991. The Influence of Graphite Stitching on the Interlaminar Performance of Composite Structural Elements. Center for Composite Materials Technical Report 91-30. Newark: University of Delaware.

Hung, Y.Y. 1989. Shearography: a novel and practical approach to nondestructive evaluation. Journal of Nondestructive Evaluation 8(2):55–68.

Hyatt, M.V., R.G. Caton, and D.T. Lovell. 1989. Advanced Materials Development in Commercial Aircraft. AIAA Paper 89-2127. Washington, D.C.: American Institute of Aeronautics and Astronautics.

IATA (International Air Transport Association). 1991. Conclusions of the IATA Questionnaire on Composite Structure Maintenance. Montreal, Canada: IATA.

Ilcewicz, L.B., P.J. Smith, T.H. Walker, and R.W. Johnson. 1990. Advanced technology commercial fuselage structure. Pp. 127–156 in the First NASA Advanced Composites Technology Conference, October 29–November 1, Seattle, Washington. Washington, D.C.: National Aeronautics and Space Administration.

Ilcewicz, L.B., P.J. Smith, and R.E. Horton. 1992. Advanced composite fuselage technology. Pp. 97–158 in the Third NASA Advanced Composites Technology Conference, June 8–11, Long Beach, California. Washington, D.C.: National Aeronautics and Space Administration.

Janicki, G. 1994. New Materials Applications for Subsonic Aircraft. Presentation to the Committee on New Materials for Advanced Civil Aircraft, National Materials Advisory Board, National Research Council, Washington, D.C., July 12.

JTEC (Japanese Technical Evaluation Center). 1994. JTEC Panel Report on Advanced Manufacturing Technology for Polymer Composite Structures in Japan. NTIS Report #PB94-161403. Baltimore, Md.: JTEC.

Kinloch, A.J. 1979. Interfacial fracture mechanical aspects of adhesive bonded joints—a review. Journal of Adhesion 10:193–219.

Kinloch, A.J., and S.O. Osiyemi. 1993. Predicting the fatigue life of adhesively-bonded joints. Journal of Adhesion 43:79–90.

Ko, F.K. 1989. Preform fiber architecture for ceramic-matrix composites. Ceramic Bulletin 68:401–414.

Kotzian, B.A. 1989. Evaluation of non-destructive testing on operators' aging aircraft. Pp. 16–17 in FAA Aging Aircraft Workshop Proceedings. Columbus, Ohio: American Society for Nondestructive Testing.

Krieger, R.B. 1984. The relation between graphite composite toughness and matrix shear stress-strain properties. P. 1570 in the 29th International SAMPE Symposium and Exhibition. Covina, Calif.: Society for the Advancement of Materials and Process Engineering.

Larson, P.H., and B.T. Åström. 1991. Design and Fabrication of an Experimental Facility for Pultrusion of Thermoplastic Composites. Center for Composite Materials Report 91-32. Newark: University of Delaware.

Lee, J.W., D.H. Allen, and C.E. Harris. 1989. Internal state variable approach for predicting stiffness reductions in fibrous laminated composites with matrix cracks. Journal of Composite Materials 23:1273–1291.

Leodolter, W. 1982. Application of Superplastic Forming/Diffusion Bonding Titanium Fabrication. Air Force Final Report AFWAL-TR-8. Wright-Patterson Air Force Base, Ohio: Air Force Wright Aeronautical Laboratory.

Lewis, W.H., W.H. Sproat, and B.W. Boisvert. 1979. A review of nondestructive inspection reliability on aircraft structure. Pp. 1–16 in the Proceedings of the 12th Symposium on Nondestructive Evaluation. Kelly Air Force Base, San Antonio, Texas: U.S. Air Force Logistics Command Program.

Marceau, J.A. 1989. Regarding corrosion control and aging aircraft. Paper presented at the 2nd Annual International Conference on Aging Aircraft, October 3–5, Baltimore, Md. Washington, D.C.: Federal Aviation Administration.

Marceau, J.A. 1994. Service Experience with Current (Aging) Fleet. Presentation to the Committee on New Materials for Advanced Civil Aircraft, National Materials Advisory Board, National Research Council, Washington, D.C., September 28.

Markus, A., and R. Palmer. 1991. Resin transfer molding for advanced composite primary aircraft structures. Pp. 271–292 in the First NASA Advanced Composites Technology Conference, October 29–November 1990, Seattle, Wash. Washington, D.C.: National Aeronautics and Space Administration.

Markus, A., P. Thrash, and B. Grossheim. 1993. Manufacturing development and requirements for stitched/RTM wing structure. Pp. 503–524 in the Fourth NASA/DoD Advanced Composites Technology Conference, June 7–11 1993, Salt Lake City, Utah. Washington, D.C.: National Aeronautics and Space Administration.

Martin, J.D., and J.E. Sumerak. 1987. Pultrusion. Pp. 533–543 in Composites, Engineered Materials Handbook, Vol. 1. Materials Park, Ohio: ASM International.

Masters, J.E., J.L. Courter, and R.E. Evans. 1986. Impact fracture and failure suppression using interleaved composites. Pp. 844–858 in the 31st International SAMPE Symposium and Exhibition, J.L. Bauer and R. Dunaetz, ed. Covina, Calif.: Society for the Advancement of Materials and Process Engineering.

McBride, S.L., and J.W. Maclachlan. 1984. Acoustic emission due to crack growth, crack face rubbing and structural noise in the CC-130 Hercules Aircraft. Journal of Acoustic Emission 3(1):1–9.

McBride, S.W., W. Boyce, J.P. Perschbacher, and W. Pless. 1987. Acoustic emission applications in the aerospace industry. Pp. 418–445 in Nondestructive Testing Handbook, Vol. 5, Sec. 12, R.K. Miller and P. McIntire, eds. Columbus, Ohio: American Society for Nondestructive Testing.

Messler, R.W., Jr. 1981. Electron beam weldability of advanced titanium alloys. Welding Journal, Welding Research Supplement 60(5):79S–84S.

Newnham, R.E. 1993. Smart, very smart, and intelligent materials. Materials Research Society Bulletin 18(4):24–25.

NRC (National Research Council). 1986. Net Shape Technology in Aerospace Structures, Vol. 1. Air Force Studies Board. Washington, D.C.: National Academy Press.

NRC (National Research Council). 1987. The Place for Thermoplastic Composites in Structural Components. National Materials Advisory Board. NMAB-434. Washington, D.C.: National Academy Press.

NRC (National Research Council). 1991. Life Prediction Methodologies for Composite Materials. National Materials Advisory Board. NMAB-460. Washington, D.C.: National Academy Press.

NRC (National Research Council). 1992. Aeronautical Technologies for the Twenty-First Century. Aeronautics and Space Engineering Board. Washington, D.C.: National Academy Press.

NRC (National Research Council). 1995a. Computer-Aided Materials Selection During Structural Design. National Materials Advisory Board. NMAB-467. Washington, D.C.: National Academy Press.

NRC (National Research Council). 1995b. Expanding the Vision of Sensor Materials. National Materials Advisory Board. NMAB-470. Washington, D.C.: National Academy Press.

O'Brádaigh, C. 1990. Analysis and Experiment in Diaphragm Forming of Continuous Fiber Reinforced Thermoplastics. Center for Composite Materials Report 90-22, University of Delaware, Newark, August.

Odagiri, N., H. Kishi, and T. Nakae. 1991. Torayca T800/3900-2 toughened epoxy prepreg system: toughening concept and mechanisms. P. 43 in Proceeding of the American Society for Composites 6th Technical Conference. Lancaster, Pa.: Technomic.

Ogo, Y. 1987. Effect of stitching on in-plane and interlaminar properties of carbon-epoxy fabric laminates. Masters thesis. University of Delaware, Newark.

OTA (Office of Technology Assessment). 1992. Green Products by Design: Choices for a Cleaner Environment. OTA-E-541. Washington, D.C.: U.S. Government Printing Office.

Pickens, J.R., F.H. Heubaum, L.S. Kramer, and K.S. Kumar. 1991. Ultra high strength weldable aluminum lithium alloys. Martin Marietta Corporation. U.S. Patent 5,032,359. July 16.

Popper, P., and R. McConnell. 1987. A new 3-D braid for integrated parts manufacture and improved delamination resistance. Pp. 92–103 in Proceedings of the 32nd International SAMPE Symposium. Covina, Calif.: Society for the Advancement of Materials and Process Engineering.

Rakestraw, M.D., M.A. Vrana, D.A. Dillard, T.C. Ward, J.G. Dillard, and C.A. Berglund. 1994. A fracture mechanics methodology for the testing and prediction of adhesive bond durability. Pp. 65–93 in Durability and Damage Tolerance, Proceedings of the 1994 International Mechanical Engineering Conference and Exposition. AD-vol. 43. New York: American Society of Mechanical Engineers.

Reifsnider, K.L. 1991a. Fatigue of Composite Materials. London: Elsevier.

Reifsnider, K.L. 1991b. Performance simulation of polymer-based composite systems. Pp. 3–26 in Durability of Polymer-Based Composite Systems for Structural Applications, A.H. Cardon and G. Verchery, eds. New York: Elsevier.

Reifsnider, K.L. 1992. Use of mechanistic life prediction methods for the design of damage tolerant composite material systems. Pp. 205–223 in ASTM STP 1157, M.R. Mitchell and O. Buck, eds. Philadelphia, Pa.: American Society for Testing and Materials.

Reifsnider, K.L., and Z. Gao. 1991. Micromechanical concepts for the estimation of property evolution and remaining life. Pp. 653–657 in Proceedings of the International Conference on Spacecraft Structures and Mechanical Testing, ESA SP-321, Noordwijk, The Netherlands, April 24–26. Noordwijk, The Netherlands: European Space Agency.

Reifsnider, K.L., and W.W. Stinchcomb. 1986. A critical element model of the residual strength and life of fatigue-loaded composite coupons. Pp. 298–313 in Composite Materials: Fatigue and Fracture. ASTM STP 907, H.T. Hahn, ed. Philadelphia, Pa.: American Society for Testing and Materials.

Richard, R., J.R. Plumer, and S.B. Driscoll. 1983. Development of a Variable Cross-Section Pultrusion (VSP) Die. Paper presented at the 38th Annual Conference of the Reinforced Plastics and Composites Institute, Society of the Plastics Industry, February 7–11.

Sachse, W., and M. Gorman. 1992. Acoustic emission measurements of aerospace materials and structures. Pp. 263–280 in Flight-Vehicle Materials, Structures, and Dynamics: Assessment and Future Directions. Vol. 4: Tribological Materials and NDE, R.L. Fusaro and J.D. Achenbach, eds. New York: American Society of Mechanical Engineers.

Sampath, S.G. 1993. Critical Research Issues Relating to Aging Aircraft. Paper presented at the Flight Safety Symposium, Zhukowsky, Russia, September. TsAGI and AGARD (Advisory Group for Aerospace Research and Development). Neuilly Sur Seine, France: North Atlantic Treaty Organization.

Seagle, S.R. 1994. Current/Future Trends in the Materials Marketplace. Presentation at the Workshop on Critical Issues for the Aerospace Specialty Materials Industrial Base, National Materials Advisory Board, National Research Council, Washington, D.C., July 13–14.

Shapery, R.A. 1990. A theory of mechanical behavior of elastic media with growing damage and other changes in structure. Journal of Mechanics and Physics of Solids 38:215–153.

Simo, J.C., and J.W. Ju. 1987. Strain- and stress-based continuum damage models. International Journal of Solids and Structures 23(7):821–839.

Skelton, J. 1989. Triaxially braided materials for composites. Pp. 117–128 in Textile Structural Composites, Composite

Material Series, vol. 3, T.W. Chou and F.K. Ko, eds. Amsterdam, The Netherlands: Elsevier.

Smith, B.W. 1994a. 777 Market Driven Materials Development. Presentation to the Committee on New Materials for Advanced Civil Aircraft, National Materials Advisory Board, National Research Council, Washington, D.C., July 12.

Smith, B.W. 1994b. Engineering Needs in Materials and Processes for Commercial Airplanes. Presentation to the Committee on New Materials for Advanced Civil Aircraft, National Materials Advisory Board, National Research Council, Washington, D.C., July 12.

Smith, P.J., L.B. Ilcewicz, and J.T. Olson. 1994. Advanced Technology Composite Fuselage. Paper presented at the Fifth NASA/DoD Advanced Composites Technology Conference, Seattle, Washington, August 22–25.

Spain, R., and C. Bailey. 1991. Apparatus and method for braiding fiber strands and stuffer fiber strands. U.S. Patent 4,984,502.

Staley, J.T. 1992. Metallurgical aspects affecting strength of heat-treatable alloy products used in the aerospace industry. Pp. 107–143 in the 3rd International Conference on Aluminum Alloys, Their Physical and Mechanical Properties, Vol. 3, L. Arnberg, O. Lohne, E. Nes, and N. Ryum, eds. Trondheim, Norway: The Norwegian Institute of Technology and SINTEF Metallurgy.

Staley, J.T. 1994. Aluminum alloys in subsonic aircraft. Paper presented at the Second Thermal Structures Conference, Charlottesville, Va. E. Thornton, ed. Washington, D.C.: American Institute of Aeronautics and Astronautics.

Staley, J.T., and Rolf, R.L. 1993. Trends in alloys for aircraft. Pp. 629–642 in Proceedings of the International Symposium on Light Metals, Light Metals Processing and Applications. Montreal: Canadian Institute of Mining, Metallurgy, and Petroleum.

Starke, E.A., and C.P. Blankenship. 1993. Aluminum-lithium alloys. Metallic Materials for Lightweight Applications, Proceedings of the 40th Sagamore Army Materials Research Conference 40:139–157.

Starke, E.A., and J.T. Staley. 1996. Application of modern aluminum alloys to aircraft. Progress in Aerospace Sciences, in press.

Steenkamer, D.A., D.J. Wilkins, and V.M. Karbhari. 1993a. Resin transfer molding I: materials and preforming. Processing of Advanced Materials 3(2):89–105.

Steenkamer, D.A., D.J. Wilkins, and V.M. Karbhari. 1993b. Resin transfer molding II: tooling and processing. Processing of Advanced Materials 3(3):181–192.

Talreja, R. 1985. A continuum mechanics characterization of damage in composite materials. Proceedings of the Royal Society of London A399:195–216.

Tenney, D.R. 1992. Aerospace applications of metals and metal-matrix composites. Pp. 1–55 in Advanced Metallics, Metal-Matrix, and Polymer Matrix Composites. Vol. 2: Flight-Vehicle Materials, Structures, and Dynamics—Assessment and Future Directions, A.K. Noor, ed. New York: American Society of Mechanical Engineers.

Thompson, D.O. 1992. Tools for a NDE engineering base. Pp. 167–180 in Flight-Vehicle Materials, Structures, and Dynamics – Assessment and Future Directions. Vol. 4: Tribological Materials and NDE, R.L. Fusaro and J.D. Achenbach, eds. New York: American Society of Mechanical Engineers.

Turnbow, M.L. 1995. Personal communication, V. Panhuise with M. Turnbow, Tennessee Valley Authority.

Vikstrom, M. 1989. Thermography of foam-core sandwich structures. Materials Evaluation 47(7):802,804.

Wagner, L., and J.K. Gregory, 1993. Improvement of mechanical behavior in Ti-3Al-8V-6Cr-4Mo-4Zr. Pp. 199–209 in Beta Titanium Alloys in the 1990s, D. Eylon, R.R. Boyer, and D.A. Koss, eds. Warrendale, Pa.: TMS.

Walker, T., L. Ilcewicz, D. Polland, and C.C. Poe, Jr. 1992. Tension fracture of laminates for transport fuselage, part 2: large notches. Paper presented at the Third ACT (Advanced Composite Technology) Conference, Long Beach, Calif., June 8–11. Seattle, Wash.: Boeing Commercial Airplane Group.

Weller, R.D. 1985. AYPEX: a new method of composite reinforcement braiding. Pp. 229–249 in 3-D Composite Materials. NASA Conference Publication #2420. Annapolis, Md: David Taylor Naval Ship Research and Development Center.

Wilson, M.L., J.D. Buckley, G.E. Dickerson, G.S. Johnson, E.C. Taylor, and E.W. Covington, III. 1989. Pultrusion process development of a graphite-reinforced polyetherimide thermoplastic composite. Journal of Thermoplastic Composite Materials 2:230–246.

Winegardner, T. 1993. Creative Pultrusions. Paper presented at the Center for Composite Materials, Composites Workshop. Newark: University of Delaware.

Xu, Y.L., and K.L. Reifsnider. 1993. Micromechanical modeling of composite compressive strength. Journal of Composite Materials 27(6):572–588.

Zawislak, S.P., and J.R. Marden. 1988. Advanced Weaving Concepts for Complex Structural Preforms. NASA Conference Publication No. 3038. Washington, D.C.: National Aeronautics and Space Administration.

Appendix

Biographical Sketches of Committee Members

JOHN A.S. GREEN, *Chair*, received his Ph.D. in electrochemistry from Queen's University in Belfast, Ireland. During his tenure at Martin Marietta Laboratories, Dr. Green worked in the areas of environmental degradation of aluminum and titanium alloys, integrated aluminum production, and new materials development. He was involved in the development of Weldalite® aluminum-lithium alloys, and XD® metal-matrix composite technology. He is currently director of Advanced Materials at Lockheed Martin Laboratories.

BERNARD BUDIANSKY received a Ph.D. in applied mathematics from Brown University. He is currently Gordon McKay Professor of structural mechanics, *Emeritus,* and Abbot and James Lawrence Professor of engineering, *Emeritus,* at Harvard University. His research has been in structural mechanics and fracture. Professor Budiansky is a member of the National Academy of Sciences and National Academy of Engineering.

DAVID J. CHELLMAN received an M.S. in materials engineering from the University of California at Los Angeles. He is currently a technical fellow at Lockheed Martin Aeronautical Systems Company. His experience is in development and application of new materials in aerospace structures, cost and weight trades in design, innovative design, and processing.

LARRY P. CLARK received a B.S. in metallurgical engineering at Washington State University. He was employed at the Air Force Wright Aeronautical Laboratories in the Manufacturing Technology program. He is currently manager of parts, materials, and processes at Boeing Defense and Space Group. His experience has been in materials processing, net-shape metals processing, and aircraft component producibility.

JOHN W. GILLESPIE, JR. received a Ph.D. in mechanical engineering from the University of Delaware. He is currently associate director of the Center for Composite Materials at the University of Delaware. Dr. Gillespie's expertise is in continuum mechanics, fracture mechanics, design and analysis of composite structures, and composite manufacturing processes.

CHARLES E. HARRIS received a Ph.D. in engineering mechanics from Virginia Polytechnic Institute and State University. He has held positions at Babcock and Wilcox and at Texas A&M University. He is currently head of the Mechanics of Materials Branch at the NASA Langley Research Center. He directs research in the area of materials characterization and the development of mechanics models of deformation, durability, and durability of aerospace materials and is the technical manager of the NASA Aging Aircraft Research Program.

MURRAY H. KUPERMAN received a B.S. in metallurgical engineering from University of California at Berkeley and an M.B.A. from the University of Santa Clara and is a licensed professional engineer in the state of California. He is currently a senior staff representative at the United Airlines Maintenance and Operations Center. His experience has been in aircraft repair and maintenance, composites and bonded structures, and aircraft operations.

PAUL A. LAGACE received a Ph.D. in aeronautics and astronautics from the Massachusetts Institute of Technology, where he is currently professor of aeronautics and astronautics and director of the Technology Laboratory for Advanced Composites. His interests are in composite mechanics, structural design, testing, and manufacturing processes.

VICKI E. PANHUISE received a Ph.D. in nuclear engineering from the University of Missouri at Columbia. She is currently director of engineering and technology at AlliedSignal Aerospace. Her interests are in nondestructive testing, advanced materials, materials testing, and artificial intelligence.

KENNETH L. REIFSNIDER received a Ph.D. in metallurgy from The Johns Hopkins University. He is currently professor of engineering mechanics and chairman of the Material Engineering Science Program at Virginia Polytechnic Institute and State University. His research is in composite fracture mechanics and nondestructive testing and evaluation.

MICHAEL P. RENIERI received a Ph.D. in engineering mechanics from Virginia Polytechnic Institute and State University. He is currently senior principal engineer in the

Conceptual Design and Structural Development Department at McDonnell Douglas Aerospace. His experience is in stress analysis, structural design, composite structures, and innovative manufacturing.

EDGAR A. STARKE received a Ph.D. in metallurgical engineering from the University of Florida. He has held positions at DuPont's Savannah River Laboratory, and at the Georgia Institute of Technology and the University of Virginia where he served as dean of engineering and applied science. He is currently the Earnest Oglesby Professor of materials science at the University of Virginia. Dr. Starke's research is in advanced alloy development, fatigue and fracture, and high-temperature aluminum alloys. He is a member of the National Materials Advisory Board of the National Research Council.

HERBERT J. WARDELL received a B.S. in engineering from Seattle University. He is currently senior staff chief at Gulfstream Aerospace. His experience is in analysis and certification of metallic and composite aircraft structures for light aircraft.